ECOLOGIES OF DISEASE CONTROL

HISTORIES AND ECOLOGIES OF HEALTH

ROBERT PECKHAM, EDITOR

ECOLOGIES OF DISEASE CONTROL

SPACES OF HEALTH SECURITY IN HISTORICAL PERSPECTIVE

**CAROLIN MEZES,
SVEN OPITZ,
AND ANDREA WIEGESHOFF**

UNIVERSITY OF PITTSBURGH PRESS

Published by the University of Pittsburgh Press, Pittsburgh, Pa., 15260
Copyright © 2025, University of Pittsburgh Press
All rights reserved
Manufactured in the United States of America
Printed on acid-free paper
10 9 8 7 6 5 4 3 2 1

Cataloging-in-Publication data is available from the Library of Congress

ISBN 13: 978-0-8229-4848-3
ISBN 10: 0-8229-4848-6

Cover art from Adobe Stock
Cover design by Melissa Dias-Mandoly

CONTENTS

ACKNOWLEDGMENTS

With the exception of the chapter by Baquero, da Silva, and Faria, all contributions to this volume were presented at the workshop "Securing Epidemic Ecologies: Disease Control in Historical Perspective," organized and hosted by the University of Marburg, Germany, in December 2021. The workshop and this edited volume were generously supported by the Collaborative Research Centre, "Dynamics of Security: Types of Security from a Historical Perspective," which is funded by the German Research Foundation (DFG). We would like to thank all workshop participants for the inspiring discussions and our contributors for their rich chapters. Annemarie Worsch and Patrick Wiesinger provided invaluable assistance in formatting the book manuscript. We would also like to thank Joana Amaral and Stephen Foose for their helpful comments and suggestions on the introduction, Judy Loeven for her editorial work, and the anonymous referees for their insightful comments, productive criticism, and encouraging feedback. Last but not least, we would like to thank Abby McAllister and Amy Sherman at the University of Pittsburgh Press and Robert Peckham, series editor of Histories and Ecologies of Health, for their encouragement and support in this project.

ECOLOGIES OF DISEASE CONTROL

ECOLOGIES OF DISEASE CONTROL

SPACES OF HEALTH SECURITY IN HISTORICAL PERSPECTIVE

Carolin Mezes, Sven Opitz, and Andrea Wiegeshoff

How has health security been shaped by ecological thinking? The COVID-19 pandemic made clear that the current regime of health security is inextricably linked to ecology through the problematizations of interspecies contacts, global microbial traffic, and environmental determinants of disease. Though often understood as a recent paradigm, this volume seeks to illuminate the much longer history of this relationship. Assembling a range of disciplinary perspectives from the social sciences and the humanities, it investigates the varying ecological orientations that have informed practices of disease control since the late eighteenth century. The case studies examine the historically situated ways in which health and security are linked in attempts to intervene in relational spatial arrangements—rendered intelligible through concepts such as environment, topology, surroundings, milieu, or ecosystem—to counter disease. By adding historical depth to contemporary articulations of health security, the volume also reassesses the recent rise and the promises of ecological thinking in social theory and the humanities.

This introduction proceeds in four steps. First, it identifies three exemplary ecological moments in the COVID-19 pandemic, when human-animal relations, population topologies, and atmospheric surroundings became matters of concern. These moments are used to demonstrate the analytical scope we've termed *ecologies of disease control* that is central to the research in this volume. With a multidisciplinary readership in mind, the next two sections introduce approaches to the relationship between disease, security, and ecology from a historical, historiographical, and social science perspective. More specifically, the second section challenges the conventional and "presentist" understanding that the securitization of health started in the 1990s. Building on this, the third section reflects on the volume's conceptual approach and discusses the use of ecological thought to analyze forms of disease control in the humanities and social sciences, sometimes referred to in terms of a "general ecology" or an "ecology of powers." Finally, the last section introduces the chapters and highlights the common threads running through the individual case studies.

COVID ECOLOGIES, OR HOW TO ANALYZE ECOLOGIES OF DISEASE CONTROL

Although governments' efforts to control COVID-19 typically offered unilinear responses to the pandemic, such as the emphasis on vaccines as "magic bullets," it was thought of and framed as an ecological problem. Practices of disease control drew on relational notions of environmental complexity and acknowledged the active role of surrounding factors at various scales. These tendencies become particularly, though not exclusively, visible in three moments that characterized health security during COVID-19: (1) more-than-human entanglements, (2) population topologies, and (3) atmospheric renderings.

First, perhaps most visibly, ecological orientations have informed our understanding of the pandemic's very beginning. To investigate where and how SARS-CoV-2 emerged, the World Health Organization (WHO), partner organizations, and member countries sent a joint expert mission to trace its origin. The aim was to investigate possible zoonotic sources of the virus, trace "intermediate hosts" along the "routes of introduction" into human populations, help prevent the establishment of other possible "zoonotic reservoirs," and limit "further emergence and transmission" (WHA 2020, 6). This ecological scheme of pathogenic emergence and evolution kept reappearing throughout the pandemic, as was the case when, for example, a mutation of the virus was detected at Dutch mink farms. In response thousands of animals were culled to avoid further spillover between animal and human bodies (Chen 2020).

This exemplifies the trope of the (re)emergence of diseases in more-than-human spaces of cohabitation, which is the main epistemological catalyst behind the so-called securitization of global health since the late 1980s. This trope, as we will discuss below, resonates with longer histories of international health collaboration and has led to considerable changes in the contemporary governance formation of global health security.

Second, during COVID-19 population security was also cast in ecological terms. The notion of "herd immunity" is a case in point. It relegates the individual body's vulnerability to the surrounding bodies' infectious status. Originally a veterinarian concept, the idea informed many national responses during the pandemic (Jones and Helmreich 2020). As the population body of the human "herd" is understood to impact individual disease susceptibility, vaccination appeared as the most promising intervention. Struggles around vaccine distribution, however, show that territorial biopolitical regimes segregate the global herd of vulnerable (human) bodies. This back-and-forth shape-shift between the governance of a global, transactional reality and segmented administrative spaces reappeared in several forms during the pandemic. Whereas the imperative to protect the networks of liberal traffic has been at the core of global health security for decades, interventions during COVID-19 cut through global supply chains and disrupted the transnational mobility assemblages (Ferhani and Rushton 2020; Gebrekidan et al. 2020; Yu and Keralis 2020). Border closings and travel restrictions constantly reconfigured possible connections.

By representational means, populations were formed in the topological spaces of dashboards, lists, diagrams, and heat maps (Everts 2020; Bowe, Simmons and Mattern 2020). They were rendered legible as segregated entities through systems of epidemiological intelligence, case reporting, and quantitative modeling. These socio-technical environments facilitated the circulation of statistical numbers in real time, as well as the simulation of possible pandemic futures. Populations were classified, redivided, and made globally comparable. In particular, the pandemic data regimes positioned populations in relation to those infrastructures deemed critical for their material subsistence. The strategy to "flatten the curve" enacted a security rationale oriented toward the continuance of "vital systems" (Collier and Lakoff 2021) for the provision of health care. During COVID-19 population security thus took the shape of an uneven and multifarious set of relations involving modes of bodily cohabitation, fractured administrative landscapes, media ecologies, and infrastructural environments. The politics of disease control operated through a population topology made up of highly heterogeneous adjacencies.

Third, COVID-19 has reintroduced ambient air into the repertoire of elemental concerns of disease control. Socially shared spaces morphed into possibly dangerous sites of atmospheric transmission, highlighting the permeability of breathing bodies to their surroundings. The finding that SARS-CoV-2 not only travels via larger respiratory droplets but also through smaller and more volatile aerosols complicated the idea of a clearly demarcated microbial "safe distance" (Li et al. 2021; Opitz 2020). Countermeasures were introduced to address this environmental aspect of pandemic space: the face mask functions as a barrier, and ventilation systems exchange and filter shared air. Atmospheric notions of disease space also informed observations that relate infection severity to other matters of aerial circulation, be it pollen exposure or particles from cigarette smoke or industrial sites, which remain in human bodies after inhalation and affect their capacities to interact with pathogens (Frontera et al. 2020; Damialis et al. 2021).

In relation to ambient air and global atmospheres, yet another dimension of COVID-19's ecological reasoning came to the fore. The pandemic reintroduced human impact on the environment as part of the problematization of health security. As the quality of water and air improved in some places due to the lockdown-enforced "anthropause," attempts were made to connect the pandemic to efforts at healing the planet from the ailments of CO_2 emission and environmental pollution (Searle, Turnbull, and Lorimer 2020). Global health governance actors demand that we "build back better," while scholars continue to urge governments to combine measures against climate change and biodiversity loss with issues of animal and planetary health. Both aim to tackle the challenges of the Anthropocene (OECD 2020; FAO 2020; Carlson, Albery and Phelan 2021).

Human-animal entanglements, population topologies, and atmospheric envelopment: these nexuses indicate how current ecological conceptualizations of infectious diseases translate into practices of health security—and vice versa, how practices of disease control frame the pandemic in terms of ecological management. Responding to SARS-CoV-2 has become a matter of dealing with the reality of being situated in, and environed by, dense webs of causative forces, a matter of regulating dynamic situations of material relationality and codependence. This is not to say that COVID-19 marks the beginning of an ecological paradigm in infectious disease control. Rather, in important respects it iterates those epidemic worldviews that have informed the so-called securitization of global health for about three decades. Responses to the pandemic, however, have brought to the fore the ecological tendencies already at work in the regime of global health security.

At the same time current modes of ecological reasoning strongly resonate with historical forms of disease control that have, in their own ways, foregrounded elemental envelopments, populations as part of a material milieu, the role of landscapes, encounters with animals, or the topologies of imperial networks, inter alia. This volume explores these resonances and puts the following question center stage: *How have epidemics been thought of, controlled, analyzed, and politically dealt with as matters of ecological relatedness?* This question presupposes a notion of ecology that understands disease as the result of complex patterns of interconnectedness, thereby challenging causal and one-directional explanations. With such a broad framing, the chapters are neither limited to situations when the term *ecology* is explicitly used, nor to the scientific development of the concept. Rather, and partly inspired by Etienne Benson's (2020) approach of studying how ideas take shape in practical settings, this volume traces how notions of diseases as relational entities have been operative in a diverse array of attempts at controlling them. Hence we do not understand ecology as a fixed concept but rather trace the multiple articulations of spatial relationality in practices of health security. The case studies excavate different forms of material interdependence, the vital role of environmental surroundings, and the entangled nature of modes of cohabitation at play in outbreak situations.

With this focus, we seek a new way of analyzing and historicizing processes that are conceived of as a "securitization" of health. The Copenhagen School in International Relations has put forward this concept in order to shed light on how security logics have spread beyond the narrow field of military affairs and how existential threats are rhetorically conjured up to justify extraordinary protective measures (Buzan, Wæver, and De Wilde 1998). This perspective is instructive to investigate how diseases have been framed as threats to national security since the late 1980s, both in terms of bioterrorism and the destabilizing potential of outbreaks in regions with weak public health infrastructure (Davies 2013; McInnes and Rushton 2013). However, securitization analysis of this sort has two limitations in capturing how the apparatus of health security places disease within an ecological framework—limitations that this volume seeks to overcome.

First, due to the "grammar" of securitizing speech acts, security in the field of health becomes associated with an adversarial logic. Focusing on securitization processes tends to highlight exceptional policies that target pathogens as inimical intruders, triggering militarized responses to protect the boundaries of healthy organisms, both individually and collectively. While this focus is insightful, it risks overlooking important features of security at work. Since it presupposes clear-cut, unilinear oppositions, it is not well equipped to conceive a rationale of security that accommodates principles of ecological complexity. Moreover, the relatively narrow view on speech acts has difficulties taking into account the spatial features of material entanglements and the technological repertoires deployed to act on them. The case studies therefore turn their attention toward ecological orientations as a hinge between conceptualizations of disease and practices of disease control to demonstrate the intricate workings of power in relational configurations of health security. This not only sharpens the critical sensibilities of security studies but also reminds us of the importance of thinking about power itself as relational.

The second shortcoming of securitization analysis lies in its historical limitations. The current regime of global health security is often said to be relatively recent, not older than three or four decades. Even though some scholars point to the long tradition of legal collaboration in international health, most descriptions of the changes since the late 1980s remain "presentist" in their understanding (Elbe 2010; McInnes and Lee 2012). In contrast, this volume sets out to historicize recent dynamics, proposing the lens of the ecological as an entry point for investigating how epidemic configurations have been problematized and rendered governable as relational enitites since the late eighteenth century. It explores how changing ideas about entanglement, networked dynamics, topological enfolding, or surroundedness were imbricated in historically diverse practices of controlling disease. The historical perspective therefore exposes a multitude of concepts and techniques that complicate germ-centered, monocausal concepts of disease (Honigsbaum and Méthot 2020; also Mendelsohn 1998; D'Abramo and Neumeyer 2020). Even in the heyday of germ theory, scientific concepts highlighted the role climate, landscape, and technology played in the interactions between living beings and microbes—a perspective deeply entwined with the (in)security of colonial expansion, as well as with the perception of societal risks of intruding into the web of life (Anderson 2004). Ecological "styles of reasoning" (Hacking 1994) are integral to frameworks of disease control that predate the emergence of ecology as a scientific discourse, but that nevertheless conceive of disease in terms of patterns of interconnectedness, socio-material environments, and spatial relatedness. Excavating those frameworks is instructive for putting the current modes of disease control into sharper relief and at the same time correcting the assumption about the relatively recent origin of health security.

AIRS, WATERS, AND PLACES, OR HOW TO HISTORICIZE ECOLOGIES OF DISEASE CONTROL

The current global health security apparatus has developed alongside the concept of emerging infectious diseases (EID), which has served as a major epistemic driver, evoking the constantly looming threat of disease, setting the "world on alert" (Weir and Mykhalovskiy 2010), and demanding preparedness for as yet unknown public health dangers (Lakoff 2007; Sanford, Polzer, and McDonough 2016). Since the 1990s an "emerging disease worldview" has come to govern threat perceptions as well as preventive and countermeasures (King 2002), a trend that intensified around the turn of the millennium, fueled by concerns about biosecurity, biological warfare, and bioterrorism (Cooper 2006; Davies 2008; Rushton 2011; Wenham 2019; Elbe 2010).

At its inception, against the backdrop of the HIV/AIDS crisis (Elbe 2009), the EID concept was a powerful rejection of post–World War II overconfident expectations about the eradication of epidemics, when antibiotics, vaccines, and pesticides seemed to promise ultimate control over infectious diseases. Amid such scientific optimism, and shortly after WHO had declared the global eradication of smallpox in 1980, the outbreak of an unknown deadly epidemic in North America, and shortly after in Europe, shocked the Western medical community and the public alike (Snowden 2008; Lederberg, Shope, and Oaks 1992; Morse 1996). HIV/AIDS can be seen as a prototype of new diseases resulting from a zoonotic spillover of pathogens between animal hosts and humans, which occurs more frequently due to closer interspecies contact in the wake of population growth, extensive land use, and changes in ecosystems (Narat et al. 2017; Keck and Lynteris 2018). As the mantra of the EID worldview has it, "in today's interconnected world, a health threat anywhere is a threat everywhere: an outbreak in a remote village can spread to major cities on all six continents in less than 36 hours" (CDC 2017). This sense of global proximity is tightly coupled with ecological renderings of infectious diseases in both the thinking about EIDs and wider global health security efforts. It also informs current projects, such as the One Health initiative, that argues against the artificial boundaries set between human and veterinarian medicine (for a critical discussion, see Wolf 2015; for a historical perspective, see Woods et al. 2017). Antimicrobial resistance and planetary health are other examples of the intertwined nature of medical theories, ecological thinking, and health security. Problematizations of antimicrobial resistance relate health to the well-being of vital microbial communities in the socio-natural environment of vulnerable bodies (Landecker 2016; Hinchliffe 2021). The discourse on "planetary health" considers the climate, the quality of the soil, or the composition of the air as crucial for human health (Dunk et al. 2019). However, from a medical history perspective, such ecological frames appear all but specific to the period since the 1990s.

Regarding the spatial rendering of global health security, it is important to note that the trope of the unprecedented speed at which epidemics spread is not new. Rather, it resonates with historical discourses about the health-related dangers of economic integration, specifically in imperial contexts. In their analysis of late nineteenth- and early twentieth-century maritime hygiene technology, Lukas Engelmann and Christos Lynteris point out that "the trope in itself is in fact the product of the turn of the nineteenth century. The fact is best illustrated by the identical nature of maps used to warn about pandemic danger, where the globe appears to be spanned by a thick web of lines; now airplane flights, then shipping routes" (Engelmann and Lynteris 2020, 12). It was along the routes of traffic, trade, war, and slavery that "the 1800s saw the greatest redistribution of pathogens the world has ever known" (Harrison 2015, 652). Imperial economies and colonial rule depended on humans, cattle, and goods moving between colonies and imperial metropoles, as well as on the productivity of enslaved and exploited populations. Therefore, to lessen economic damage caused by heterogeneous and disparate quarantine regulations and, simultaneously, to address the issue of increasingly global epidemics, attempts at international agreements were made as early as 1851 at the first International Sanitary Conference. Four decades later, in 1892 the participating states of the seventh conference adopted the first International Sanitary Convention (Huber 2006; Harrison 2006; de Almeida 2015). Whereas the politico-juridical framework changed considerably with the EID concept, collaborative efforts at health security date back much further.

Indeed, concepts that link health to factors outside the body and in turn instruct efforts of disease control, also have a longer history. The chapters in this volume focus on this nexus as it materialized in the past, but also more recently in moments of communal response to epidemic outbreaks. However, the case studies do not simply project contemporary concepts onto historical cases to identify forerunners of current thinking or "securitizing moves" but heed the caution of historians, such as Mark Harrison, who warn against ahistorically reducing the past to a teleological prehistory of the present (Harrison 2017). With this in mind, we investigate how relational understandings of disease took shape in historically situated practices of disease control by analyzing how such practices were constitutively attuned to ideas about environmental complexity, the active role of surroundings, and material interdependencies between more-than-human agencies. The chapters consider changing ideas of relatedness and spatial situatedness by engaging with case-specific settings, such as the built environment, infrastructural conditions, human and more-than-human disease reservoirs, underlying metabolic conditions in the body, landscapes, and climates. In doing so, distinct and sometimes conflicting notions of (in)security come into view that concern not only questions of health but also, for instance, economic or political interests.

With respect to the deeper history of ecologies of disease control, two aspects appear crucial. First, and closely connected with Harrison's warning, narrowing the focus to prominent features of current debates runs the risk of missing important parts of the story. As the following chapters show, there is a veritable multitude of ecological orientations at work in historical projects of disease control. The relationship between climate and race is a case in point. Especially from the eighteenth century on, geographically and increasingly biologically deterministic notions of racial differences informed medical assumptions about susceptibility to diseases that became immensely influential in the history of colonial expansion and slavery (Chakrabarti 2014, 57–70; Nash 2014). Expanded historical perspective is therefore an opportunity to identify the enduring political and medical legacies of such racialized concepts in ecological thinking (Anderson 2006; Harrison 1999).

Second, we are well advised not to overstate continuities or, as Conevery Bolton Valencius (2000, 24) puts it, to mistake a "family resemblance" for a "shared history." A historicizing perspective can reveal any traditions that strongly resonate with recent configurations of infectious disease control, but also highlight discontinuities and disruptions. Hippocratic thinking about *Airs, Waters, and Places*, an important reference point to approach the environmental dimension of medical history (e.g., Valencius 2002; Jankovic 2010), simultaneously underscores and illustrates this aspect. Advocating a holistic perspective on health and disease, the ancient physician Hippocrates and his disciples focused on the relationship between human bodies and their surroundings. In this view environmental influences played a critical role in causing diseases by disturbing the bodily balance of "humors." Though refined, amended, and challenged, these concepts remained critical to medical thought and practice well into the eighteenth and nineteenth centuries, not least because they proved to be "remarkably resilient and adaptable" (Bashford and Tracy 2012, 495; see also Cantor 2002). Moreover, as Charles Rosenberg reminds us, the "emphasis on the body as always situated and always in process, always interacting with the environment that sustains and threatens it" (2012, 668) connects Hippocratic views even with current medical concepts.

Regardless of these continuities, over time crucial shifts have taken place. Not only has our understanding of the human body and its boundaries and relationship with the environment changed, as ideas about the (inter)dependency between climate and humans exemplify (Bashford and Tracy 2012), but also the medical thinking about environmental surroundings. In this respect, interpretations of the environment as a potential "harbourer of pathogens" (Valencius 2000, 20), following the rise of bacteriology, were not identical to the older emphasis on airs, waters, and places, even though they similarly stressed the relevance of environmental factors for human health.

Around the turn from the eighteenth to the nineteenth century— the chronological starting point of our volume—a radical break with ancient medical authorities occurred in Europe as diseases were situated in a socio-technical urban milieu. With the emergence of clinical medicine, especially in Paris during and after the French Revolution, the hospital became the main place of knowledge production. Clinical physicians not only firmly established the thinking of diseases as discrete phenomena; they also began to study statistical correlations between the occurrence of diseases and environmental factors, such as foul odors. These accounts directly influenced the powerful sanitary movement that emerged in Britain during the 1830s and 1840s (Snowden 2020, 168–87). The advocates of sanitary reform promoted a strictly localist understanding of infectious diseases, attributing them to filth and the resulting corruption of the surrounding air. Illness appeared less a result of poverty and problems created, or worsened, by urbanization and industrialization; rather, public health issues seemed solvable through sanitation programs addressing water supply, drainage, sewerage, cleanup campaigns, and personal cleanliness. This approach presented authorities with a straightforward technical fix for social issues (Berridge 2016, 42–60; Hamlin 1998; Kiechle 2017).

Despite their focus on what we now might call an infrastructural ecology, the rather one-directional, causal relation between filth and disease propagated by the sanitary movement was considerably less complex than later ecological concepts of mutual entanglements. Within this narrower framework, the sanitarians nevertheless touched on concerns that are important within a shared history of ecological orientations in disease control. A case in point is the example of human-animal relations, which are today at the heart of efforts to protect health security. The sanitarians drew attention to dangerous interspecies proximities that, in their view, exposed humans to noxious smells in urban spaces (Kirk and Worboys 2013, 566). The problem of disease transmission across species began to occupy medical research from the mid-nineteenth century on (Hardy 2003, 201; also Cassidy 2019; Keck 2019). With the advent of germ theories, animals became increasingly seen as pathogen carriers or "epidemic villains" (Lynteris 2019), and therefore the targets of countermeasures. Historically, interrelations with nonhuman animals have been problematized in various ways under the aegis of health protection. For instance, in the early 1900s the knowledge that rat fleas were disease vectors of the plague led to a veritable war on rats aimed at combating and preventing outbreaks (Dyl 2006; Wiegeshoff 2021). Threatening intimacy between species resulted not only from urbanization processes that brought human and nonhuman animals into closer contact. Interspecies relations were also changed by the large-scale interventions into ecosystems that characterize the history of colonialism, imperialism, and capitalism, demonstrating the need to understand today's spillover scenarios within their longer history (Sivasundaram 2020, 296–97; see also Beinart and Hughes 2007; Ross 2017).

The fact that research increasingly focuses on such aspects of the history of health and disease points to certain recent historiographical trends. Historical studies on the environmental and ecological dimensions of medical history have gained considerable steam since the early 2000s. Although questions of place and space characterized the work of medical historians in the United States and Europe as early as in the 1930s and 1940s (Valencius 2000, 3–5), the "Hippocratic turn in medical history" (Sellers 2013, 450) is a relatively recent development. Over the last twenty years, studies have shown where and how relational notions of space continued to matter in medical sciences (e.g., Honigsbaum and Méthot 2020; Honigsbaum 2016; Tilley 2004; Jones 2004). In the late twentieth and early twenty-first centuries this focus of inquiry was inspired particularly by the discipline of disease ecology, as well as the EID framework (Sellers 2013). It is unlikely a coincidence that renowned historians in this field, such as Warwick Anderson, Chris Sellers, and Susan Jones, also have a background in the medical and veterinary sciences.

Moreover, environmental history, a field that has prospered since the 1990s, has begun to influence the history of medicine and decenter human agency in the process (Nash 2014; Mitman, Murphy, and Sellers 2004). Such perspectives found their way into analyses concerned with the history of infectious diseases (e.g., Nash 2006; Fressoz 2012; Honigsbaum 2020; Snowden 2020; Crosby 1972; McNeill 1976; McNeill 2010). However, much is still to be done to integrate environmental, ecological, and medical perspectives into historical research, not least in conceptual and analytical terms (Alagona et al. 2020; Sellers 2018; Green 2020; Diener 2021; Otter, Breyfogle, and Brooke 2015). With this volume we contribute to this discussion by bringing together perspectives on ecologically oriented practices of disease control from historical, sociological, anthropological, and human geographic ap-

proaches. But as Linda Nash (2015) reminds us, although we benefi-
cially draw inspiration from today's scientific concepts to frame more-
than-human histories, it is crucial to understand these very concepts
within their own historical moment, shaped by (geo)political, social,
and cultural developments. We reflect on this analytical challenge in
the next section.

CRITICAL TENSIONS BETWEEN GENERAL ECOLOGY AND ENVIRONMENTALITY

Today, we argue, the ecological appears in two distinct ways: on the one
hand, we have identified ecological orientations immanent to the field
of health security; on the other, ecological frameworks have attracted
scholars from various disciplines, dissolving any phenomenon into a set
of constitutive relations favored as a non-reductionist analytical strategy.
In order to qualify this overlap between the empirical and the conceptual,
this section delineates how ecological modes of inquiry have become the
linchpin of critical analyses in the social sciences and humanities. We
then discuss the Foucauldian notion of "environmentality" as exemplary
for navigating the current tensions of the ecological, recognizing it as
integral to health security *and* adopting its critical analytical capacities.

A turn to ecology as a conceptual, if not ontological frame is epito-
mized in Bruno Latour's call not to modernize but to "ecologize" (Latour
2007). According to Latour (2017, 220–55), such reorientation is necessary to render visible the vital ties of the "earthbound" to their surroundings. Subscribing to an even stronger premise of entanglement, Donna
Haraway (2016, 49) stresses the ethical implications embedded in the
"webbed ecologies" of collective life. By dissolving being into "multi-
species-becoming-with," Haraway's ontology of "co-existence" is inseparable from the invocation to cultivate mutual "response-ability" (Haraway 2016, 63, 34). Her playful suggestion to rename the humanities
into "humusities" (in the sense of humus) captures the ecological reorientation she has in mind. Both Haraway and Latour were inspired by
Isabelle Stengers's (2005) account of an "ecology of practices." Primarily concerned with scientific knowledge production, Stengers (2010, 32)
points to the "ecological situation" of material interdependencies that
operate in "truth telling" events. Still other authors, such as Jane Bennett
(2010), stipulate that the ecological framework extends from organic life
to inorganic matter of all kinds. Her "political ecology of things" cir-
cumscribes the "agentic assemblages" of minerals, debris, fatty acids, or
electricity grids. In this way, ecology becomes the universal key to a vital
materialism (Bennett 2010, 107).

Modes of earthbound existence, more-than-human ethics, knowledge production, inorganic things (this list could go on)—in each case, the turn to ecology is part of a wide-ranging conceptual move. First, ecology is taken beyond the concerns of traditional environmentalism to traverse the boundaries between nature and culture, society and the elemental, life and technology. In this sense ecology morphs into "general ecology" (Hörl 2017, 7–8) to study the transversal lines running between environmental, societal, and affective "registers" operating in parallel (Guattari 2000, 28). Second, the approaches pit the ecological framework against a particular blend of social constructivism. The dissolution of any given entity into a heterogeneous network of relations not only highlights the complexities of interdependence and the shortcomings of causal explanations, but also promises a step beyond "culturalist" accounts that focus too narrowly on language, meaning, or symbolic orders (Coole and Frost 2010). This (re)turn to the material dimensions of the social in the early twenty-first century points beyond a purely academic discourse as it is closely connected to scientific findings about the Anthropocene which underscore the limitations of human agency and human dependency on natural forces and resources. Under these conditions, the critical gesture of pointing to the socially constructed nature of seemingly fixed entities has lost touch with the agentic properties of the world, and has therefore "run out of steam" (Latour 2004, 225). The orientation toward neighborly associations, entanglements, and assemblages is intended to open novel avenues for critique.

This generalized ecological framework has been productively applied to health security. For instance, Stephen Collier, Andrew Lakoff, and Christopher Kelty asked authors to contribute to a special issue on the investigation of "Ebola's ecologies": "The concept of 'disease ecology' typically refers to a pathogen's relationship to a natural milieu—particularly animal hosts and their environmental niche—and to how this milieu is affected by human behavior. Here, however, we conceive of Ebola's ecologies more broadly to include the administrative, technical, political, and social relationships through which disease outbreaks evolve, and into which experts and officials are now trying to intervene in anticipation of future outbreaks" (Lakoff, Collier, and Kelty 2015). In this way they extend the ecological analysis to trace the multiple elements involved in making up Ebola across the divide of culture and nature. Hence, Ebola's ecologies are more-than-human without being less social or less political.

Another example is Steve Hinchliffe and colleagues' (2016, 54–67) collaborative work on biosecurity in livestock farming, in which they propose the topological concept of the "disease situation" to articulate similar ecological sensibilities. In order to avoid linking disease to the causative agent of a pathogen or contaminant, they propose to think of "pathogenicity" as a potential that is inherent to particular disease situations, understood as configurations of "intra-active" entities. From this view, bacterial foodborne diseases emerge from a web of heterogeneous relations that reach across specific sites: "The way labour practices intra-act with poultry guts, . . . or changing farming practices intra-act with pig bodies and microbes, . . . or the way local authority budgets intra-act with food safety inspections . . . all affect the disease potential of the situation" (Hinchliffe et al. 2016, 15). This general ecology of disease situations is invested with a decisively critical edge. By adding economic schemes, nutritional habits, or logistical regimes to the repertoire of pathogenic actors, it provides an analytical vantage point on the material conditions of disease emergence.

Ecology thus seems good to think with, especially in attempts to provide a material account of health security without sticking to the biomedical model and its reductionism (for a general critique see King 2017, 22–27). Yet what happens to the critical aspirations of this conceptual strategy when, as outlined above, ecological orientations become an organizing feature of health security itself? The historical gaze reveals that the complicity of ecological orientations with structures of power and exploitation runs deep, as ecological modes of reasoning often function as a lever for, at times, highly invasive control practices and interventions. This does not in itself limit the analytical potential of ecological thinking but it reminds us of the legacies it carries with it and invites further critical inquiry. We are, of course, far from being the first to address such a ghostly relation, in which the empirical material haunts the analytical apparatus. Most notably, and almost paradigmatically, this is at the core of Michel Foucault's notion of an "environmental" security dispositif. His otherwise well-rehearsed take on biopolitics offers insight into maneuvering the doubled appearance of the ecological.

Dealing with the governmental response to smallpox, Foucault points out a proto-ecological form of spatial ordering at the core of the liberal security dispositif. Since the late eighteenth century liberal security has tied the life of populations not only to the flow of resources and commodities, but also to the local climate, vegetation, and weather. Instead of directly targeting individuals, it reckons with the risks emerging from the material forces of the milieu in which collective life exists and "circulation is carried out" (Foucault 2007, 20). At one point in his analysis, Foucault (2008, 261) even changes the notion of *gouvernementalité* into *environmentalité* to qualify the shift in the political logic: liberal security presupposes that "environmental spaces" operate "autonomously" and therefore resorts to the "regulation of environmental effects."

Carolin Mezes, Sven Opitz, and Andrea Wiegeshoff

Over the last decades the concept of environmentality has been used to shed light on the widespread insertion of "eco-knowledges" into biopolitical calculations, especially during the second half of the twentieth century (Luke 1995, 69). Considering populations as "biologically bound to the materiality within which they live" (Foucault 2007, 21) has intensified security concerns reckoning with a "generalized crisis environment" (Massumi 2009, 155). Across a "spectrum of threat" (154) that ranges from war to weather, environments appear as reservoirs that incubate highly uncertain and sudden events of emergence triggering emergencies (Cooper 2006). Accordingly, "environmental power" does not seek to protect the environment. It treats it as a complex, unpredictable, and potentially catastrophic force field to be reckoned with. In this way environmentality is always already immersed in an "ecology of powers" (Massumi 2009, 173).

Many studies have observed how such modes of power play out in current practices of health security and epidemic control. Against the backdrop of the rediscovery of infectious diseases in the medical sciences and the EID concept, scholars examine how the governmentality of health security conceives of disease threats in terms of a "microbial traffic" (Morse 1992, 1362) that is entangled with a variety of circulatory processes (Barker 2015; Voelkner 2011). Various studies demonstrate how WHO's International Health Regulations relate the control of disease events to the immanent potential of a wider ecology of mobile elements. In this ecology human movement is closely connected to the mobility of vital materialities such as containers, food, water, or animals (Opitz 2016). Further, the referent object of health security—what is deemed worthy of protection—shifts from the individual body and the population to the resilience of wider socio-technical systems considered vital (Lakoff 2017, Mezes 2024). At the same time the aim of mapping "the highways and bridges of viral traffic" also directed the focus of public health authorities to "landscape drivers of disease emergence" (Fearnley 2020, 61, 58). In an attempt to forestall the emergence of avian influenza, a "turn to eco-virology" (54) has taken place that employs remote sensing technologies to identify "host networks" (57). Already during the Ebola crisis of 2014–2015, the tele-epidemiological gaze from above, via satellite technologies, harbored the promise of spotting "ecological 'fault-lines' where disease might arise" (Peckham and Sinha 2017, 29). These examples exhibit the multiple ways in which practices of health security operate in an environmental mode when it comes to monitoring and controlling disease outbreaks. Even the most recent treatment of the microbiome is cast in terms of a "probiotic environmentality" (Lorimer 2017, 34). Other than traditional hygiene practices, new forms of environmental probiotic management try to foster health through the production of new vital ties.

The notion of environmentality thus has proven useful in showing that a "general-ecological relationism . . . is inscribed, to a certain extent, within the history of control" (Hörl 2017, 7–8). It is therefore indicative of the problem at the core of this volume. Through specific case studies, we learn what "ecological relationism" in disease control consists of, and how environmentalities unfold in situated practices of health security. In and with this focus, the ecological appears as an overdetermined political epistemology. The observation that there are two ecological orientations, namely one in practices of health control and disease security and another in recent social science and humanities' analyses, demands a delicate maneuver: the capacity to critically conceive of ecological orientations within the field of health security while at the same time subscribing to a relational model of power.

CONTRIBUTIONS TO THE DEBATE

To translate our initial observations about a nexus between ecology, health, and security into a broader interdisciplinary conversation, we deliberately sought to include a diverse range of approaches to empirical material, methodology, theorizing, and conceptual work. The result was an authors' conference at the end of 2021, with scholars at different career stages and from backgrounds in history, sociology, human geography, and cultural anthropology, as well as contributions from interdisciplinary research networks including veterinarian and entomological expertise. Overall, the history of the publication was itself influenced by the new and intermittent ecologies of disease control imposed by COVID-19, albeit locally in different ways. These experiences made their way into some of the case studies that follow.

In terms of time and space of specific ecologies of health security, the contributions cover case studies from the late eighteenth century until the present, on topics ranging from disease-ridden slave plantations in the British Caribbean to the COVID-19 crisis. We thus concentrate on the period during which the idea of diseases as distinct entities became firmly established in the medical sciences and, at the same time, epidemics turned truly global in their geographical reach. It was in the long nineteenth century that epidemic diseases started to affect all inhabited continents (Harrison 2017). Since then medical history has seen multiple efforts and attempts to control what was perceived, time and again, as almost unstoppable outbreaks. Such time-, place-, and group-specific problematizations of infectious diseases reflect different understandings of disease-inducing factors and, in turn, of effective countermeasures.

They also involve a large variety of spatial configurations with their respective means of knowing.

Whereas each chapter makes its own case about the relationship between disease, security, and ecological orientations, read together they contribute important insights to the very multi- and interdisciplinary debate the volume wishes to start. Each highlights different ways in which forms of relational thinking in disease control were tied to power and, in turn, time and again stabilized societal orders within their hegemonic structures. Some of the cases show how control practices concerned with environmental relations and more-than-human ecologies selectively excluded aspects such as poverty or social vulnerability, and thereby in fact reduced complexity and ignored sociopolitical factors. Other contributions critize the promises of contemporary ecological concepts in health security such as One Health. Further still, in several case studies conceptual or methodological suggestions are made to reframe ecologies of disease control. Yet others re-narrate a generic account of a modern era of reductionism in health and disease control. While many of the chapters speak to more than just one aspect, the structure of the volume follows three organizing themes: first, the observation that understanding ecologies of disease control entails analyzing them as ecologies of power; second, the insight that knowing disease ecologies does not necessarily entail ecological modes of knowing; and third, that conceptualizing forms of ecological relatedness through and by way of case materials is a productive analytical approach.

Ecologies of Disease Control as Ecologies of Power

This section deciphers ecologies of disease control as ecologies of power. Ecologically oriented practices of disease control are not necessarily an alternative to hegemonic forms of power and knowledge, but they are often shaped by them. They may undergird structures of oppression, exploitation, deprivation, or neglect, and divert governmental attention away from inequities. Thus they may be used to impose and perpetuate socioeconomic assymetries, be it of individuals or social classes.

Susan Jones's case study on disease control in Central Asia during the 1920s and 1940s underscores this point and simultaneously adds layers to the early history of disease ecology. The Soviet lens of her case reveals a strand of epidemiology that largely avoided reductionist bacteriological approaches. Soviet disease ecologists studied ecological interactions and focused their countermeasures on disrupting these connections. Crucially, such efforts on the ground formed an essential part of governmental strategies to control land and people in the Soviet hinterlands and along its borders. However, not only did the local environments of the Central

Asian steppe prove resilient and hard to control; struggles and conflicts around interventions highlight rather different, often competing notions of security in relation to health. From the government's perspective, unruly people and unhealthy landscapes endangered the stability of the regime, border security, and productivity, whereas local resistance shows that there were different needs and threat perceptions. In this context problematizations of ecologies of disease unveil the heterogeneous character as well as the limits of territorial orderings.

Karina Turmann's work turns to late eighteenth-century control of "the yaws" on British slave plantations in the Caribbean. In this case the historical perspective opens a variety of spatial problematizations and imaginaries of environmental control, which are infused by extremely reductionist, racist accounts of human life and disease. Considered a noncontagious disease caused by unhealthy environments, efforts to control yaws were closely intertwined with attempts to secure the reproduction of slave populations. Together with the growing abolitionist movement, such diseases posed an imminent threat to the economic welfare of planters. Turmann shows how European medical practitioners framed their control measures around the highly racialized, enslaved maternal body as an object and the enslaved women as a threat to their newborn's health. Her account of the various spaces where notions of dangerous environments formed at the turn of the century also includes the trope of the island climate. It highlights the allegedly unhealthy entanglements of food and racialized bodies in the "tropics," the problem of lodgings and built environments in specific climates, and the socio-spatial stratifications of the plantation estate.

Focusing on the research on pellagra in the southern states of the United States in the early 1900s, Julia Engelschalt demonstrates how US public health experts framed a disease caused by malnutrition as an insect-borne "disease of place" and in turn closely monitored local environmental conditions in affected regions. American officials effectively downplayed the crucial role poverty played, thus legitimizing specific environmental approaches, such as vector eradication, while drawing attention away from the underlying social problems. Moreover, as Engelschalt demonstrates, such frameworks proved to be compatible with germ-centered theories of disease and reductionist approaches to disease control. Her analysis challenges conventional narratives about ecological approaches as successful ways to more fully capture the complex realities of outbreaks.

Moving to very recent history, Oswaldo Santos Baquero, Sara Cristina Aparecida da Silva, and Júlia Amorim Faria provide insight on the ecologies of violence emerging in and at the same time coproducing health emergencies, as in the COVID-19 pandemic. Building on the notion of multispecies health and ethnographical observations from community research on the living conditions in a São Paulo favela, their analysis points out a specific more-than-human landscape of syndemic violence which is shaped by "marginalizing apparatuses." They challenge overly optimistic narratives about the critical reach of One Health. Stressing the role of violence in the urban periphery and thus highlighting insecurity, precarity, and vulnerability, the chapter critically reframes the question of ecology in analyses of health security and disease control.

Knowing Ecologies?

The second group of case studies demonstrates that practices of "knowing ecologies of disease" are by no means practices of "ecological knowing." Looking at innovations in diagnostics, at topological understandings of diseases in epidemiological models, or at systems of syndromic disease surveillance, the case studies trace important conceptual shifts in health security and public health. But they also show that knowledge practices emphasizing the relational quality of disease do not necessarily produce more comprehensive or integrated accounts. Rather, across modern history they often tended to reduce relational complexity or introduce reductionist accounts, albeit in different disguises.

In his chapter on disease modeling between the 1880s and the 1920s, Lukas Engelmann traces the history of a technique that holds a central place in ecologically oriented notions of disease causation and control as well as in epidemiological knowledge production. Discussing a crucial shift in epidemic theory, in which space was reconfigured from a topographical environment into a set of topological vectors, Engelmann demonstrates how the modeling of infectious diseases, on the one hand, rejected a causal or deterministic influence of the environment while turning it, on the other hand, into a uniform and abstracted variable. As a result, disease models inspired visions of controlling and eradicating diseases, such as malaria, through universally applicable measures. Rather than engaging with different features of the natural or built environment, early modelers developed theories in which epidemics were, to a certain degree, conceptualized independently from place and time—a moment of dis-environmentalization inherent to the spatialized account of disease in modeling.

Focusing on a more recent development, Henning Füller investigates "syndromic surveillance" as it emerged in the United States. Based on ethnographic material, his chapter outlines how the surveillance of outbreak events—one of the core moments in the "securitization" of health—has developed since the early 1990s. Füller extrapolates on how this specific "way of knowing" (Pickstone 2001, 1), by virtue of its infrastructural basis, enfolds a twofold problematization of the ecological. On the one hand, the surveillance problem of situational awareness is turned into an environmental problem for infection control, as the infrastructurally created "baseline" normalizes specific contexts and becomes an abstract topological milieu for event detection. On the other hand, the chapter analyzes how the idea of One Health is deployed in syndromic surveillance, rendering it a rather hollow form of cross-sector holism. His critical assessment of One Health in practice resonates with Baquero's, da Silva's, and Faria's account.

Finally, Steve Hinchliffe's chapter critically engages with the promise of the emergence of new diagnostic tools intended to improve knowledge and treatment of disease in lifestock farming. Building on the social science of classification and diagnoses, he argues that recent shifts in livestock farming toward on-farm and data-intensive systems may, paradoxically, perpetuate the very threats they seek to allay. In the formulation of the "diagnostic machine," diagnosis is understood as a style of knowing that is not only concerned with making disease present, but also helping to consolidate and perform a set of social, material, and living relationships—not least relations of a capitalization of food production and marketization of health devices. Despite their promissory security logic, diagnostics may do little to ameliorate the ecological and health insecurities associated with livestock systems. They allow perpetuation of a mode of production that has been—due to its tendencies toward simplification and densification, medicalization and exploitation—at the center of health security concerns for years.

Thinking with and Rethinking Ecologies of Disease Control

The case studies in this section develop an analytical sensibility for the "doubling" of the ecological moment described above: its appearance in health security as well as in the analytical accounts thereof. Folding the empirical into the conceptual and vice versa, the chapters put forward new ways of thinking the ecological in disease control. They understand, for instance, the biological discourse on metabolism as a resource for situating the planetary condition of the Anthropocene within the human body; they address infrastructure not only as a research site but as a frame for conceiving of entanglement; or they take the risks of breathing as a model for elemental relatedness. In this sense the contributions in this final section oscillate between thinking *about* and thinking *with* ecologies of disease control.

Ann Kelly and Clare Herrick's chapter turns to contemporary interventions into disease ecologies. They analyze three spatial rationales that organized control efforts during Ebola outbreaks: the ring, the reservoir, and the frontier. Their case provides an example of the recent paradigm shift in global health security, in which so-called emergency R&D—the accelerated development, approval, and deployment of medical countermeasures—has become a cornerstone of emergency response and governance. In a study on vaccination trials during outbreaks, they track the highly fraught politics of inclusion in Ebola immunization and lay bare a repertoire of three spatial "infra-logics" on which the biopolitics of emergency R&D relies. With the practices of ring vaccination, the efforts to secure the human viral reservoir, and the attempt to push the health emergency frontier in R&D, new ecologies of global health knowledge and action emerge that aim to balance humanitarian demands caused by a public health care crisis, commercial investment in medical countermeasures, and the realities of supply shortage. Through and from the empirical material, their analysis develops the aforementioned three spatial figures that can guide critical analyses of complex and complicated relations in disease control.

In their contribution Uli Beisel and Carsten Wergin highlight the hybrid entanglement of mosquitoes, people, landscapes, and transportation systems linked to the ecological transformations that are responsible for the spread of vector-borne diseases. They draw on the notion of "transecology" to rethink current approaches to health security in these domains. Focusing on how *Aedes* mosquitoes—which transmit diseases such as dengue, West Nile, and yellow fever—figure as invasive species, Beisel and Wergin argue that, in addition to warming climates, mobility infrastructures are key actants in shaping disease ecologies. The authors introduce the ethnographic method of "infrastructural go-along" to delineate such complex, technologically saturated "transecologies." Ethnographic knowledge not only complicates the notion of "invasions" as it is to be found in health security discourse; it also complements and corrects the visual knowledge provided through public health maps that spatialize disease risks emanating from modes of being with mosquitoes in particular ways.

Hannah Landecker takes the notion of ecology into the human body. Her chapter shows how SARS-CoV-2 exposed the extent to which metabolic disorders form an "underlying condition" that increases the risk for a more severe course of the disease. In Landecker's account, the underlying condition reveals a relational biology that is thoroughly anthropogenic: air pollution or industrialized diets have disrupted the fragile interactions between mucus fluids and microbiota that secure bodily surfaces against viral entry; chronic inflammations provoked through exposure to endocrine-disrupting chemicals have put the body in a constant state of alarm, thereby ultimately exhausting its defense mechanisms. This particular ecological rendering of the body leads to a refined understanding of the Anthropocene: it is not to be found on some higher, planetary scale but registers in the metabolic milieu, an event that is "also taking place at the gut wall." With this radical change in perspective, Landecker abducts matters of disease control from hegemonic practices of preparedness planning, disease detection, and outbreak response. She ties them instead to the ecological dynamics that secure or disturb the maintenance of bodily boundaries.

The final case study focuses on health's elemental dimension from a related, yet different angle. Sven Opitz analyzes how air, as a transmission medium for virus-laden aerosols, has brought ecologies of breath to the fore in our lives, but not yet to that of analysis. For attuning sociology to this elemental condition, the chapter puts forward the concept of the "atmosocial" and elaborates three features: First, with its voluminous extent, the atmosocial exceeds "territories of the self" (Goffman 1971, 28) and challenges their ordering principles. Second, it conjoins the turbulent fluid dynamics of respiratory life with the affective dynamics of highly uncertain atmospheric encounters. Third, its cloudy texture denotes a cohabitation of bodies that do not interrelate as clearly demarcated entities. In materially sharing intimacies of breath, they are enmeshed in their milieu. The atmosocial thus differs from modes of relatedness more familiar in sociology, such as interactions or networks, and points to a form of ecological entanglement that does away with the idea of a clearly demarcated *Umwelt*.

The book closes with a commentary by Melanie Kiechle. As a historian, she explores the question of change over time and structures her observations along the question of authority and power, the role of technology, and social relations as they run through the case studies. In doing so, Kiechle reflects on the difficulties of thinking with and through ecology, particularly paying attention to long-term impacts of disease events.

PART I

ECOLOGIES OF DISEASE CONTROL AS ECOLOGIES OF POWER

CHAPTER 1

SECURING THE STEPPES

SOVIET SCIENTIFIC KNOWLEDGE, ENVIRONMENT, AND DISEASE, 1920S–1940S

Susan D. Jones

What did *security* mean during the decades in which Josef Stalin and his supporters consolidated their control over the Soviet Union? For the Stalinist Communist Party leadership,[1] an important meaning of *security* was controlling land (including plants and animals), people, and productivity in its far-flung hinterlands and border regions. This chapter analyzes the roles of biomedical scientists in creating frameworks of disease surveillance and control, which functioned as an integral part of incorporating the peoples and landscapes of Central Asia into the Soviet Union and ensuring state security in these hinterlands. Specifically, Soviet biomedical scientists developed an influential model of epidemiology that supported collectivization and its attempt to dominate and regulate people, animals, and landscapes to make them productive for the Soviet state.[2] Using the case study of disease control in parts of Central Asia, the Kirghiz/Kazakh grasslands and semiarid steppes, especially in the 1920s and 1930s, this chapter briefly analyzes how the local environment and the work of scientists, physicians, and technicians shaped the central government's efforts at securing this vast region.

Productivity was essential to the Stalinist goal of rapid industrialization, and every region and republic of the USSR were expected to contribute, from the inauguration of the first Five-Year Plan in 1928. To feed the burgeoning numbers of industrial workers during the 1930s and on, the hinterlands would be required to raise ever-larger numbers of food-producing plants and animals. These hinterlands included the vast steppes of Central Asia, encompassing the semiarid region along the northeastern side of the Caspian Sea that stretches east to Lake Balkhash (today part of the Republic of Kazakhstan).[3] After the Russian Revolution and civil war, the Soviet central leadership sought to continue aggressive tsarist-era policies by further integrating the homelands of the Kazakhs (whom Russians erroneously called "Kirghiz") into the Soviet Union. To make the steppes into a breadbasket, hundreds of Russian workers were to be relocated there to become farmers (continuing a long history of Russian settlement from tsarist times) (Seitz 2019, 181). More imported workers, along with local labor, built the infrastructure for another important institution of state control: the Karaganda Corrective Labor Camp (Karlag), a large and notorious gulag (Barnes 2011; Hedeler and Stark 2007). The development of Karlag proceeded alongside plans for agricultural and industrial development and the intended erasure of nomadism on the Kazakh steppe.

Nomadic peoples in Kazakhstan would be forced to abandon their lifeways and become sedentary farmers whose labor and animals were compelled to serve the central government's plans. This desire to erase nomadism was not new; tsarist Russia had spent considerable resources on controlling Tatars, "Kirghiz," and other nomadic peoples. But the events of the late 1920s and early 1930s were cataclysmic. Food shortages, drought, and the violence associated with forced collectivization created a perfect storm in which millions of Central Asians died, particularly in Kazakhstan. The crisis was so severe that a postindependence Kazakh presidential commission and several scholars have called it a "genocide" or a "crime against humanity" on the semiarid steppes.[4] For Kazakhs living through the famine and the violence between 1929 and 1935, the Russian vision of their lands as a breadbasket was a bitter irony.

For Soviet central planners, the steppes were hostile lands populated by "backward" peoples whom they viewed as ignorant, unsanitary, and unproductive. These people were also intractable and resistant: the 1916 rebellion across the steppes had been the most recent serious challenge to Russian/Soviet control, and Moscow had not forgotten it. The steppe people who survived it had not forgotten, either; their rebellion was brutally repressed, followed by famine and war.[5] Both sides were wary of the other. This region was also considered a security risk because part of the Kazakh steppe bordered the Soviets' most powerful neighbor, China. Until 1949 China was nominally led by a republican government, but vast swaths of the country were held by warlords, and the border was not secure. Kazakhs regularly traveled over the border, as they had for some time, to work, visit relatives, or relocate. These movements of people, like the annual nomadic peregrinations, were difficult to predict or restrict and were a source of insecurity for governments on both sides of the border (Ohayan 2006).

Soviet central planners imagined the landscape itself as not only unproductive but also potentially dangerous. The Central Asian steppes (and forests and mountains) were viewed as unhealthy places for workers, military personnel, and settlers, and this suspicion was reinforced by recurrent disease outbreaks among imported Russian workers. In response, teams of physicians, parasitologists, ecologists, and other life scientists (whom I will call disease ecologists) were sent, under the auspices of the Commissariat for Health (Narkomzdrav),[6] on lengthy expeditions to assess the conditions and make recommendations for controlling the steppe, its diseases, and its peoples. Between 1930 and 1950 two new institutional systems, the Kazakh People's Commissariat for Public Health (Kaznarkomzdrav) and the Antiplague Institute network, were constructed to assert permanent control over disease problems. Narkomzdrav and its subsidiary branches deployed scientists, physicians, technicians, and public health officials to assess, transform, and maintain a secure hold over intransigent landscapes and peoples in the steppelands. Moscow officials expected these scientists to exert control over nature, subordinate people into central planning, and secure the borders of territory recently incorporated into the Soviet Union. In turn, scientists, local people, and the local environment influenced the results of central government planning on the ground (Michaels 2003).

THE SOVIET LENS

Focusing on the Soviet context provides fresh perspectives within the multiple Western scholarly traditions that frame this story: the history of Cold War science within a more-than-human theoretical framework, the concerns of disease surveillance versus individual autonomy, and theories of state formation and maintenance. Scholars have begun to incorporate the "more-than-human" approach, which for our purposes can be defined as the entangled roles of both human and nonhuman actors (such as climatic effects or animals) in creating historical changes over time (Pyyhtinen 2016, chap. 4; Whatmore 2006). Nonhuman actors have agency independent of humans, and they often escape or frustrate human control; in this way, they are active participants in dynamic assemblages of relationships. They contribute to and constrain transformations. In this case the Soviet transformation of the steppe into a controlled, productive, secure region was a process of *becoming* that could not be isolated from the processes of the changing seasons, the cycles of the indigenous animals and their diseases, and the lifeways of the steppe peoples. Securing the steppe against disease (as a barrier to productivity) did not always turn out the way Soviet central planners envisioned, despite their development of powerful systems of biomedical assessment and surveillance.[7]

The Soviet case also reframes our ideas of surveillance, control, and security. Western scholars have characterized systems of disease surveillance and control in opposition to cultural values of individual autonomy and privacy. Fairchild, Bayer, and Colgrove (2007) proposed an influential framework of analysis that assessed the impacts of disease surveillance in tension with "privacy, a value central to American democratic life" (Fairchild, Bayer, and Colgrove 2007, xviii). Not surprisingly, the Soviet Union and its public health apparatus has been viewed by many American commentators as the antithesis to the American model.[8] Implicit in this critique was a lack of privacy, freedom, and autonomy for Soviet citizens that would be unacceptable in the United States. But as we will see, the situation on the ground in the USSR, even in highly controlled areas, was far more complex. In southern and eastern Kazakhstan the Western notion of "privacy" was not the major source of tension with state surveillance. Instead, a powerful cultural system of tribal autonomy and landscape use, which was the source of security and survival for the local people, resisted the often violent incursions of the tsarist and Soviet states. For example, in Kazakhstan local leaders and intellectuals participated in governance and state formation, shaping Kazakh self-determination and the terms of Kazakh incorporation into the Soviet Union despite the terrible famines, dislocations, and destruction of the early Stalinist years (Amanzholova 2004).

Finally, framing surveillance in terms of landscape transformation is not unfamiliar to Western scholars. The Soviet collectivization case study even appeared in James C. Scott's classic *Seeing Like a State* (1998), in which he argued that collectivization was an example of "high modernism . . . on a leveled social terrain" whose planners had the advantage of "starting from zero" in an empire ravaged by war, revolution, famine, and economic collapse by 1920. Scott's focus on "order" as the principle underlying the young Soviet state's "revolution in culture," and the new industrial model of collective agriculture, works for the Central Asian

Susan D. Jones

borderlands. But collectivization was not built on a "bulldozed site" of ruin; the local situation was far more complex than that. Durable community-based indigenous governance and resource-use systems had long enabled people to live in the harsh steppe climate (and made them more resistant to state control) (Scott 1998, 193–96).[9] Rather, Scott's argument that statecraft sought to transform "the population, space and nature . . . into closed systems that can best be observed and controlled" applies here (82).

Between the late 1920s and early 1950s (the period of "high Stalinism"), the Soviet state reacted to a sense of insecurity about its frontier Central Asian regions by creating ways of knowing and controlling that included scientific assessment of the local disease environment and technical work within biomedical institutions. However, this process was messy, complex, and often did not achieve central planners' goals. Far from Moscow, Russian and Central Asian scientists, physicians, and technicians; local community leaders; and local people all acted according to their own needs and desires. These actors, and the more-than-human elements of the environment (e.g., climate, soils, plants, animals), often challenged Soviet efforts to secure the Central Asian frontiers.

SOVIET SCIENTISTS AS THE EYES AND EARS OF MOSCOW

In the late 1920s and early 1930s the Central Asian hinterlands remained somewhat mysterious to central planners in Moscow. Biomedical scientists and technicians were deployed to assess the barriers to productivity on the steppes, deserts, and taigas, especially the periodic outbreaks of diseases such as bubonic plague, malaria, and encephalitis that sickened and killed people working there. Those scientists working specifically on bubonic plague, who spent years in the hinterlands investigating the sources of this dangerous disease, called themselves *chumalog* (plagueologists) (Khudyakov and Suchkov 1999, 4–13).[10] Since most did not specialize in just one disease, this chapter will refer to biomedical scientists working in the hinterlands more broadly as "disease ecologists." Most were trained in the biological sciences (ecology, parasitology, zoology, entomology) or as physicians; some held degrees in multiple disciplines. They usually traveled as members of multidisciplinary expeditions to places remote from the centers of scientific learning in St. Petersburg and Moscow. These disease ecologists, and their theories and practices, were at the center of Russian epidemiology during the Soviet decades.[11]

Soviet epidemiology developed very differently from that of the West; it placed much more emphasis on environmental causes and conditions and ecological frameworks of disease, eschewing the more strictly reductionist bacteriological approach familiar to Western historians (Jones and Amramina 2018). While still using microbiological techniques, Soviet disease ecologists combined them with surveys of the microorganisms, insects, and animals in the natural environment. This broader ecological approach, they argued, revealed the ultimate sources of diseases that spilled over and took hold in human populations. Despite its importance, the Soviet ecological approach to epidemiology has not received much attention outside Russia and the former Soviet bloc. Until very recently Anglophone histories of the Soviet biological and biomedical sciences have mainly focused on the ascendency of the idealogue Trofim Denisovitch Lysenko and the subsequent tragic persecution of geneticists and other biologists whom he viewed as his competitors.[12] However, despite the dangerous years of Lysenkoism (the mid-1930s and 1940s), many biologists flourished while working in disease ecology. This was not an impoverished scientific practice, but a rich one, based in environments that otherwise remained inscrutable and uncontrollable to the Soviet Central Committee in the 1930s–early 1950s. Scientists were the eyes and ears of Moscow, but they were also able to be reasonably independent in terms of pursuing interesting research questions and interacting closely with local people.

Scientific ways of knowing, and cadres of scientifically trained people, were key tools that the young Soviet regime inherited from the nineteenth century.[13] The tsarist-era Russian centers of training, overseen by the Academy of Sciences, produced generations of scientists, and there was already a long tradition of Russian scientific expeditions to the hinterlands. Expeditions were supposed to gather as much information about exploitable natural resources (including people) as possible. Harnessing this scientific tradition in the service of the state made sense because it supposedly served all stakeholders' interests: the central government, the scientists themselves, and the people they were supposed to be "improving." An excellent example is the group of disease ecologists, pictured in figure 1.1, who participated in Central Asian expeditions led by Academician Evgeny Nikanorovich Pavlovsky. In 1931 this group also included Polina Andreevna Petrishcheva, her Turkmen assistant Mengli Davletov, and several trainee disease ecologists, entomologists, and physicians. Between 1928 and 1940 Pavlovsky made one or two expeditions each year to Central Asia. From 1930 Petrishcheva was stationed in Kara Kala, Turkmenia (today's independent nation of Turkmenistan). She lived there for years and studied local diseases, such as malaria, visceral and cutaneous leishmaniasis, nematode diseases, and spirochetosis. Her specific interest in the ecology of insect vectors in the region allowed her to gather large volumes of data on mosquitoes, fleas, and ticks and connect these data to epidemic outbreaks of malaria,

FIG. 1.1. Evgeny Nikanorovich Pavlovsky, Polina Andreevna Petrishcheva, and members of the parasitological expedition, Turkmenistan 1931. Source: ФОФ-66244, Military Medical Museum, St. Petersburg, Russian Federation. Courtesy of the Military Medical Museum.

recurrent typhus, and leishmaniasis.[14] Petrishcheva greatly influenced Pavlovsky's ideas, and she and Pavlovsky remained close colleagues until his death in 1965.

Building on the ideas of other scientists and field experiences, Pavlovsky published a series of articles and books that made his reputation as the "father" of the most important twentieth-century Soviet disease theory: the theory of transmissible diseases originating in natural foci (природная очаговость трансмиссивных болезней, hereafter "the natural focus theory") (Jones and Amramina 2018; Korenberg 2018).[15] "Natural foci" were the deserts, steppes, caves, and forests that functioned as "biocenoses," ecological systems that supported the microorganisms that caused diseases. These diseases, he argued, arose from intricate ecological interactions between climate factors and landscapes where the animals and insects resided. Plague, for example, required burrowing rodents to incubate the plague bacilli and parasites (fleas) to spread the infection between the animals. Fleas thrived in the moist, cool environment of the rodents' burrows. Huge colonies of gerbils and jirds (small burrowing rodents) on the steppes harbored plague continuously at endemic levels, with occasional outbreaks that commonly eradicated whole colonies of the animals. Humans living or working in the vicinity could also become infected, leading to a plague epidemic (Jones and Amramina 2018).

Other epidemics in the hinterlands could likewise be traced to animals and insects in the environment: encephalitis viruses spread by ticks and malaria and arboviruses spread by mosquitoes, for example. Rabbit and marmot hunters could also become infected with tularemia or plague when skinning the animals and preparing the meat. Therefore disease outbreaks threatened the productivity of workers sent to the hinterlands to build infrastructure. Disease ecologists' mandate was to search for, identify, and make recommendations on how to eradicate these potential disease outbreaks. At the same time Pavlovsky and his students were engaged in the time-consuming scientific process of data collection and analysis, looking for underlying patterns. Pavlovsky later recalled the process as "linking separate, disparate facts, observations and experimental results that . . . [when the time is right] develop into a theoretical generalization . . . which ultimately grows into the main provisions of a real law [of nature]" (Pavlovsky 1961, 10–11).

Official reports and archival evidence from Pavlovsky's expeditions link these disease ecologists' research with the push for collectivization and the timing of Central Asian regions' incorporation into the USSR. Piecing together Pavlovsky's travels, we see a pattern of how he and his colleagues served as Moscow's eyes and ears during the 1920s and 1930s—the era of national delimitation in the USSR.[16] Most of Russian Turkestan (Central Asia) became officially incorporated into the Soviet Union in stages between about 1920 and 1936—a time in which Pavlovsky participated in yearly expeditions to this region. For example, Pavlovsky traveled and conducted research in 1928, 1930, and 1931 in Kapa-Kara (today's Kyrgyzstan) (Pavlovsky 1928–1932).[17] Established on October 14, 1924, as the Kara-Kirghiz Autonomous Oblast of the Russian Soviet Federative Socialist Republic, this area was drawn under Soviet central control in 1926 when it was transformed into parts of the Kirghiz and Kazakh Autonomous Soviet Socialist Republics (ASSR). At the time Pavlovsky was working there, the Kirghiz ASSR was being assessed for its resources and potential as an agricultural cash-crop region before the 1936 Soviet Constitution made it a constituent republic of the USSR.

The scientists had to ascertain which disease problems were already present in the region and which ones were likely to spread or intensify with irrigation, cultivation, and livestock raising. In the Kazakh SSR the taiga harbored tick encephalitis, plague was endemic in the steppe gerbils and montane marmots, and irrigating the semiarid lands threatened to spread malaria. Pavlovsky's correspondence with Petrishcheva and other colleagues detailed the complexity of the work. Equipment for recording data (Pavlovsky was an early adopter of film technology), microscopes, traps and cages for animals, and all other supplies had to be brought in by railroad, wagon, and cart.[18] Depending on local people for knowledge and labor, the scientists collected ticks, fleas, and mosquitoes, searching for infectious vectors of disease. They trapped rodents, dug out rodent burrows, and walked or rode for miles surveying the extent of rodents' colonies. They set up makeshift laboratories and developed special protective clothing to protect themselves against infection. These activities contributed to a vast trove of knowledge about Central Asian landscapes and were also the basis for the natural focus theory. In short, Soviet theories of disease causation and control were built on the process of rendering Central Asian lands, plants, animals, and people *knowable* and *controllable*.

The natural focus theory contributed to the process of collectivization by arguing that agricultural development was beneficial to the landscape and local people. Pavlovsky's argument was that first, disease arose from specific ecologies; therefore the way to combat disease was to disrupt the ecological interactions within the natural focus. This disruption could include removing one of the crucial participants in the disease's ecosystem (such as a host reservoir animal or the insect vector carrying the disease). However, the most effective disruption would be the harshest: the "sanitation" or "disinfection" of the landscape, in which all potentially disease-carrying animals and insects would be annihilated, native plants removed by burning, and finally the soil turned by plowing. The disease would then lose its natural focus and would no longer be transmissible into human populations. In support of this position, Pavlovsky cited the results of an early antiplague campaign. "We can assume that in the distant past the focus of [bubonic] plague was much larger" in the area around the Caspian Sea, he wrote. "Increased farming efforts . . . ploughing up areas of virgin land, the appearance of villages, and elimination of marmots and gerbils" had significantly decreased plague outbreaks (Pavlovsky 1966 [1965], 187–88, 205). This new high-modernist ecology of the sanitized landscape represented success for local and central public health officials, but scientists on the ground were aware that vast areas with intricate ecologies remained in Central Asia to be studied. They continued to develop the natural focus theory and discuss its ecological complexity even as they sent reassuring official reports back to Moscow.

Pavlovsky and his colleagues continued to make versions of this argument in official publications and reports throughout the 1930s from different Central Asian regions, and they specifically named collectivization as the remedy for diseased landscapes. Pavlovsky asserted that the agricultural development and settlement associated with collectivization were both the means and the goal of controlling diseases originating in natural foci. In 1933 the Council of People's Commissars of the Tajik SSR (Sovnarkom)[19] was concerned enough about an outbreak of a mysterious "paralyzing disease" to fund a series of costly expeditions undertaken by Pavlovsky and his colleagues. They identified the disease as tick encephalitis, caused by viruses that circulated among rodents, insects, and the encroaching human populations. While this was a major scientific discovery, the value of this research was characterized by Pavlovsky in his official report as "the unity of theory and practice: . . . the main objective at this stage [is] to give assistance to collective farm development" (Pavlovsky 1937). Pavlovsky's reports thus carefully articulated what he thought officials of Sovnarkom, Narkomzdrav, and the central leadership in Moscow expected to hear: that collectivization was supported by scientific as well as economic studies. At the same time Pavlovsky claimed scientific priority for elucidating disease systems and demonstrated the practical uses of the natural focus theory to the state's goals.

THE SOVIET DISEASE SURVEILLANCE SYSTEM'S ROLE IN TRANSFORMING AND MAINTAINING CONTROL OF KAZAKH HEALTH

Pavlovsky and Petrishcheva did more than report back to Moscow in the service of assessing areas for collectivization. They spent time and resources on public health activities that they viewed as their duty to the people living in these regions. They did not officially report most of these activities. These two scientists felt that they "lived quite closely" with the people of this area and that they had humanitarian reasons to help them (Petrishcheva 1960, 22). Although not mandated to do so, Pavlovsky and Petrishcheva worked to improve living conditions and prevent disease outbreaks. They helped to build and clean up water supplies, distribute crude mosquito netting, and educate people about disease prevention. For example, Pavlovsky, also a military physician, had the idea of creating posters about how malaria was spread. He had them translated into local languages and posted in villages and trading centers as part of an educational effort (Lebedenko 1961, 25, 94–95). He also examined patients suffering from tick encephalitis and malaria and attempted to treat them (Pavlovsky 1966 [1965], 22–23). This work could be viewed as self-serving or merely protecting workers as state assets, but Pavlovsky's reticence in official reports argues against these as his primary motivations. If, in the course of his scientific work, he had the time and space to contribute to local peoples' well-being, he did so. This does not excuse his contributions to the destruction of Central Asian peoples' lifeways and autonomy. Rather, these unreported activities support a view on the part of the scientists that the official, scientific, and local humanitarian work were not inconsistent with each other. The complexity of the work reflected the complexity of local peoples and environments—and their interactions with the scientists.

These activities represented a transition for scientists and physicians during the mid-1930s. Although exploratory expeditions to identify obstacles to central planning goals continued until interrupted by the Great Patriotic War (World War II), the role of physicians and scientists in the developing hinterlands was shifting over time as Soviet infrastructure accumulated. They were increasingly being sent to the steppe as clinical employees for the new hospitals and institutes being rapidly built to support industrial and agricultural regions. Central planning included the expansion of medical and public health infrastructure and infectious disease surveillance systems. As Paula Michaels has argued, the Soviet-era medical schools, hospitals, clinics, and field stations were meant to function as modernizing influences. Medical personnel were to be the "face of the regime," to model the ideal Soviet citizen's behavior and ideological beliefs (and dissuade the traditional medical beliefs and practices of the local people) (Michaels 2003, 103). As the medical system grew alongside the collective farms and industrial centers, outbreaks of the diseases studied by Pavlovsky, Petrishcheva, and the other disease ecologists continued to occur. Particularly alarming were reports that bubonic plague had reappeared in the Nagorno-Karabakh region of the Azerbaijan SSR in 1930 and was spreading around the Caspian Sea. If plague jumped across the Caspian, it would encounter the very favorable ecological conditions of the Kirghiz steppe, with its vast colonies of wild burrowing rodents that could spread plague widely.

Between 1930 and 1950, two institutional systems, the Kazakh People's Commissariat for Public Health (Kaznarkomzdrav) and the Antiplague Institute network, were constructed to assert control over infectious disease problems. The Kaznarkomzdrav, as Paula Michaels has argued, was explicitly charged with assessing the social conditions among Kazakhs that would make them vulnerable to infectious diseases—especially by equating nomadic lifeways with capitalism. The superior Soviet way was to abolish nomadic lifeways and maintain modern surveillance and control over workers' lives and bodies to ensure good health (Michaels 2003, 55; Starks 2008). Healthy workers were crucial to meeting agricultural production goals, but health surveillance had the added effect of sorting people into categories of concern: the vulnerable, the weak, and the noncompliant. Places and landscapes could also be categorized as unhealthy or dangerous, and this belief tied medical surveillance to the surveillance of landscapes. Surveilling landscapes for potential disease outbreaks became the job of the antiplague system, which grew dramatically between the late 1920s and 1950s.

The Soviet regime inherited a nascent Antiplague Institute system (API) from its tsarist predecessor, which included medical observation outposts along the border with China. The mother institute of these field stations was planned for Saratov, on the Volga River and near an area known for plague outbreaks (this institute is known today as Mikrob) (Melikishvili 2006, 30). During the mid-twentieth century the API system grew to include a core of six major research institutes and dozens of regional and field stations in or near disease foci. The API system incorporated many existing field stations and even university faculties, and it provided the training for disease ecologists (especially those working with dangerous microorganisms) throughout the Soviet period. Existing scientific disease research units in eastern Russia (at Irkutsk, Vladivostok, and Khabarovsk) and Central Asia (Ashkhabad, Turkmenistan; and Frunze, today's capital of Bishkek, in Kyrgyzstan) became official API stations in the 1930s, while those in Kazakhstan became part of the API system during and after the Great Patriotic War (ben Ouagrham-Gormley 2006, 36–39). Many of these stations were located along the troubled border: in Kazakhstan, for example, at Taldy-Kurgan in the east and Shymkent in the south. The Dzhambul observation station, reporting to the Shymkent regional station, was also in the region that served as a major destination for peoples from around the USSR who had been subjected to internal exile from the 1930s on. Finally, as Sonia ben Ouagrham-Gormley has noted, disease outbreaks were not the only reason to establish surveillance stations along the USSR's borders: the API stations bordering the Islamic nation of Iran, for example, were "mainly motivated by a false premise" that plague and other diseases originated there (ben Ouagrham-Gormley 2006, 36). Pavlovsky was sent to Iran in 1942, where he and his colleagues (mostly military) scouted water supplies, disease problems, and settlements along the border (working alongside wary British military personnel) (Pavlovsky, 1942–1943). It is important to note that a major source of Soviet authorities' sense of insecurity in broader Central Asia came from the notion that Islam was a source of resistance and organized rebellion against the regime (Amanzholova 2009; Khalid 2015). This notion's contribution to the establishment of border surveillance stations in places like Turkmenistan is unclear and deserves further archival research, but it suggests a complex interrelationship between disease surveillance and human population surveillance in these regions.

The landscape surveillance and disease control work conducted by the API scientists and workers adhered to an official Soviet policy developed in the antiplague campaigns of the 1930s: the disinfection and sanitation of landscapes. The public health terminology reflected both the role of Narkomzdrav in funding and supervising the API system and the intellectual connection between human health and landscape health. A sanitized landscape was a secure landscape, free from the ecological systems that harbored dangerous diseases. API scientists supervised the killing of wild rodents (using poison placed in the burrows) and insects (using DDT and other chemicals) and removal of the animals' food sources by burning or applying herbicides to native plants (Jones et al. 2019). Using a grid system developed by planners at the Mikrob API in Saratov, API scientists stationed in Central Asia divided up this vast region into targeted surveillance territories (this system is still used today) (KSCQZD 2010). The "labor of disinfection," to borrow a term from historian of medicine Jacob Steere-Williams, was mostly done by local people, often women and children from the families of collective farms and industrial workers. This effectively enlisted people in the destruction of their own lands and drew them more tightly into the web of central planning and control (Steere-Williams 2019).

Between 1954 and 1956 the API system leaders successfully petitioned Moscow for large amounts of money and other resources to mount a final assault on bubonic plague in its endemic foci. Not surprisingly, the central and regional Ministries of Agriculture and Health, along with the Republic party administrations, supported and contributed to this effort to remove native vegetation and animals (Pavlovsky 1966 [1965], 205). The human inhabitants of the natural nidi were also subject to control: over 20 million people were vaccinated against plague on a mandatory basis with live, polyvalent plague vaccine (supposedly with few side effects) (Pastukhov 1959, 10). When a natural nidus had experienced no cases of plague for a length of time, it was said to be sterilized. Scientifically, sterilization meant few or no cases of the disease in the local humans; few or no burrowing animals; and the inability to isolate the disease-causing microorganisms from the environment (Pavlovsky 1966 [1965], 206–8). Despite this unprecedented effort to destroy the natural foci across Central Asia, plague, encephalitis, and other natural focal diseases persisted. To be sure, the numbers of humans infected by these diseases fell dramatically during and after these campaigns. But the ultimate goal—sterilization of the steppes and eradication of the diseases—eluded the API scientists and their patrons in Moscow, Saratov, Alma-Ata, and elsewhere.

The Central Asian natural foci, and their animal, plant, and microorganism components, were active participants that could escape human control and shape the activities of settlement, collectivization, and industrialization. Even disinfected landscapes could recover their old ecologies. If even a few wild rodents stubbornly persisted on the steppe, the population grew, migrated, and created new colonies. These new colonies in turn created new opportunities for the circulation of microorganisms and vectors, and pockets of plague persisting in these animals could be transmitted to humans. For example, API officials were shocked when the supposedly disinfected Southern Volga region suffered a major outbreak of bubonic plague in 1945. Even after the 1954–1956 eradication campaigns, plague broke out again in Central Asia in 1958, 1965 (Turkmenistan), 1962, and 1966 (Kyzyl Orda region, near Kazalinsk in Kazakhstan).[20] For Pavlovsky, the natural foci predated human incursions and could exist independent of human activities; he undoubtedly understood that complete eradication was next to impossible. However, he and the other disease ecologists chose to emphasize the successes of efforts to control natural foci of disease as measured by the reduced numbers of infected people. While this was a major goal of the disease eradication campaigns, it was also essential to their own careers that scientists and public health officials communicate the most positive interpretation of their results.

DISEASE CONTROL AND HEALTH SECURITY THROUGH THE SOVIET LENS

Understanding and controlling the epidemic ecologies of the Central Asian territories contributed to multiple meanings of *security* for Soviet planners: food security, border security, security against diseases, and control of workers and the natural environment. Over the years, the problems had changed—from subduing recalcitrant nomads to harvesting more bushels of wheat, for example—but analyzing the work of a small group of disease ecologists demonstrates how crucial the biomedical sciences remained to Soviet efforts to secure the steppe. Twenty years of working to convert and invigorate the Transvolga Region and parts of the Central Asian steppe to "ensure high and stable harvests" underlay the grandiose program of Stalin's final years, the Transformation of Nature plan (1948).[21] It was in this context that the work of the disease ecologists can be seen as participating in a broader environmental agenda, but it also reveals the limits of political control.

Just as diseases like bubonic plague and tick encephalitis proved resistant to control, so did the scientists who studied them and the biocenose (the ecosystem in a particular location) that was their home. While the disease ecologists' work, as reported to Moscow, lent scientific authority to the idea that nature could be remade according to the Soviet imaginary, their attitudes were far more complex, as expressed in their scientific and personal communications. Disease ecologists were certainly practical. They resembled agriculturalists: if an animal was designated a pest, or a landscape in need of sanitation, they supported its transformation. But they also recognized that the ecological integrity of landscapes would suffer. Pavlovsky, for example, warned about the unintended consequences of ecological interventions, such as the toxicity of DDT. Insect resistance to DDT increased in direct proportion to the chemical's use, creating the opposite of the desired effect. The "tangential destruction" of beneficial insects (such as bees and other pollinators) or others necessary for a healthy ecological system could cause "collapse of the biocenosis" (Pavlovsky 1966 [1965], 202–3). Pavlovsky's biocenose was not static, but a dynamic and reactive system that needed to be understood to be managed, controlled, and made ever more productive. However, these scientists' subjects of study—landscapes with complex ecologies—proved to be actors in shaping the scientists' opportunities and epistemologies. The biocenose could (and did) evade control and respond to human intervention in unexpected ways, forcing the scientists to shape their theories to accommodate the agency of the nonhuman world.[22]

Even after Stalin's death in 1952, the Soviet economy continued to be structured around central planning with set productivity quotas for the hinterlands. If anything, the pressure on the steppe to be productive in terms of crops and livestock intensified, for example with Khrushchev's Virgin Lands campaign (1954–1960), which increased the ploughed area of the USSR by forty-six million hectares. The API system's great disease eradication campaign of the mid-1950s was meant to help agricultural productivity—and to expand disease surveillance on the steppes and borders. This surveillance applied to people as well as landscapes. By the 1980s the Soviet disease surveillance infrastructure was among the world's largest, covering some three hundred million people in the USSR (Wuhib et al. 2002, 2).

Applying the Soviet lens provides new perspectives on the definition and deployment of notions about health security. While folded into the broader concerns of state security, health security was also a matter of environmental control and surveillance. This reflected a mutual reinforcement of Moscow's production goals and the natural focus theory, the major Soviet-era doctrine of infectious diseases, that was based on the material ecology of disease systems. The natural focus doctrine's close connection to the animals, climate, and insects living in specific locations contrasts with developments in the epidemiology of Europe and the United States, which (as Lukas Engelmann's chapter argues) was moving toward a more placeless, modeled representation of infectious diseases during this exact period. The Soviet case also provides a useful example of how human and nonhuman actors cocreated material disease systems, theories about infectious disease causation, and resistances to political control.

Finally, Soviet infectious disease history during the COVID-19 pandemic provides many warnings about visions of diseased people and landscapes and the ongoing material consequences of human incursions into, and destruction of, the natural environment. As highlighted in the introduction to this volume, COVID-19 is another disease that probably spilled over from wildlife-human interactions and is thus an ecological problem. But it has also revealed the social and political limits imposed on epidemiological ideas, processes, and technologies, from modeling to vaccines. Perhaps we cannot afford to ignore the more-than-human situating of disease after all.

CHAPTER 2

PLANTATIONS AND "THE YAWS"

SECURING ENSLAVED MOTHERHOOD IN THE LATE EIGHTEENTH-CENTURY BRITISH CARIBBEAN?

Karina Turmann

In 1801 a manager of Sir Gillies Payne's plantation on the island of St. Christopher (now St. Kitts) was said to have reported that in the 1790s enslaved mothers frequently consulted him because their children were "infected with the yaws," a disease that "covers with dry scabby sores, the person who is infected with it. When cured, it often leaves him a cripple, and to children it frequently proves fatal; yet among children it generally prevails" (Caines 1801, 158). According to his evaluation of the situation, the women deliberately passed the disease on to their children, without going into detail about how the transmission occurred. He claimed that, because mothers whose children fell ill were allowed to stay away from work during their child's convalescence to care for them, they would purposefully infect them. Insisting the women were taking advantage of this situation, he locked them up with their children so that they could not pursue any activity other than healing (Caines 1801, 158–59). This manager is only one example of a "man on the spot" whose claims led his medical and nonmedical contemporaries to believe that a certain disease was widespread on plantations in the Caribbean colonies and was caused or exacerbated by the behavioral patterns of enslaved mothers.

At the time, before germ theory existed, there was no scientific evidence for the transmission of diseases and European scientific discussion shifted between "contagionist" and "anti-contagionist" theories, as well as concepts in between. One example is the concept of "contingent contagionism," that had existed since the Middle Ages but became popular only in the early nineteenth century; it implied that certain diseases were contagious only in unhealthy environments but not in healthy ones. Disease control measures accordingly focused on environmental aspects, which included not only the surroundings but also people's interaction with them or artificially created spaces. The measures reflect a reaction to the way these interactions have been portrayed, based on the physical and mental characteristics attributed to the people concerned. At the same time medical interpretations claimed that these features in turn would originate in unhealthy environments, a term usually meaning unknown or unfamiliar regions, such as the tropics in colonial contexts.

This chapter aims to contribute to perspectives on disease control in its historical manifestation and application by British colonists in the West Indies. It investigates how the local and social meanings of environment became crucial in the operational management and discursive construction of a disease and, moreover, human characteristics within the frame of spreading and controlling what was perceived around 1800 to be an "epidemic." Sources from the Lesser Antilles will be used as an empirical basis. They show how the conditions in smaller colonies varied, but also that politics, economic management, and not least medical facilities were implemented differently than in bigger colonies such as Jamaica (see Veit et al. 2023) and yet in line with larger ideologies of imperial expansion. The practices and discourses of disease control under scrutiny here focused specifically on the questions of motherhood and the health of enslaved children. These were conducted in various ways; on small plantations they were sometimes rather individually and economically motivated, depending on whether, for example, there was a strong connection or commitment to Britain or what influence British political discourse and decisions had in the colonies.

The disease assumed to be rampant on Payne's plantation, now taxonomically classified under the terms *yaws* or *frambesia tropica*, is known today as an infectious disease caused by the bacterium *Treponema pallidum pertenue*, which is transmitted by contact infection, often in the early stages of disease progression and in early childhood.[1] More rarely, however, it is also transmitted through insect bites, for example from meat flies (Kickbusch and Franz 2017, 19). Symptomatically, the disease manifests itself in the form of papules that contain the pathogen, appearing in the beginning most often on the lower legs and followed by more papules or later ulcers on the whole body. These decay after several months, often leaving scars. The later noninfectious stage (>5 years) is characterized by changes in the bones, mostly the nose and the long tubular bones, leading to deformities. Ultimately, hyperkeratosis occurs with painful cracks in the hands and feet (Schneider 2021; WHO n.d.).

From today's perspective, it may be surprising to see, on the one hand, how closely the descriptions from colonial times correspond to the symptomatology and even spaces of transmission that are now subsumed under the diagnosis of yaws. On the other hand, the sheer arbitrariness in the use of the term makes it difficult to embed the historical sources. The frequent use of the term *yaws* in reports from the British plantation colonies was probably due to its use by the enslaved people, who had their own prophylaxis and therapies for the illness (Schiebinger 2017, 62). Most of the medical terms used by colonial agents then often served as a collective term for certain types of diseases, as their causes were not clear. Better-known examples of these terminological discrepancies are in febrile diseases such as malaria or yellow fever. Nevertheless, bioarcheologic studies verify that treponema infections were common on Caribbean plantations (see Shuler 2011). Thus it will be shown why the distinct diagnosis of yaws recurrently appears in the colonial sources, and how diagnoses and indications were influenced by more than just scientific or medical factors.

The colonial explanations are consistent with the medical knowledge at the time, which was significantly shaped by the teachings of miasmatic infection theory. The theory that a disease can be transmitted via miasma persisted—among other theories that already believed in human-to-human contagion—until the nineteenth century and claimed that infections were caused by putrefactive matters in the air and water. Colonists had different understandings of infections depending on their professional background; many of those responsible for containing diseases among the enslaved people on plantations were not trained physicians. Their prevention, diagnosis, and treatment ranged among contemporary Eurocentric medical knowledge, experience, and the observation of epidemic developments on the spot. The body of knowledge on yaws in the late eighteenth century was shaped by these factors and assumptions at a time when yaws and syphilis were believed to be the "same disease, often referred to collectively as the great pox," and having "its origins in the sexual excesses of tropical men and women" (Paugh 2014, 226), evidencing racially biased perceptions of its incidence.

As the example of the Payne plantation shows, measures of disease detection and control focused on various environmental aspects that variously affected epidemic courses and the lives of the enslaved people. From an observation of symptoms, colonists inferred different diseases, which led to the many convoluted conclusions regarding its spread and containment. Today it is known that the non-venereal treponematoses infest mainly within the skin, mucosa, and bones. In contrast, the late stages of venereal syphilis damage internal organs in particular, including the cardiovascular and the nervous systems. But despite modern diagnostic procedures, even today a morphological and serological differentiation between yaws and other *Treponema* species is not possible (Krause 2021).[2] Although frambesia is not a venereal disease, due to its relationship and similar symptomatology with, for example, syphilis or leprosy, it is mentioned not only in contemporary discourses but also in the context of historical research on British slavery in the West Indies, particularly under the aspect of venereal diseases (see, e.g., Block 2017; Burnard and Follett 2012; Paugh 2014).

Yaws in the context of the politics of childbearing has been thoroughly studied by Katherine Paugh (2013, 2014, 2017). She reminds us that "historians of medicine widely agree that disease categories . . . are culturally constructed categories that are constantly being shaped and reshaped, by healers, by the sick, and also by those who write, talk, and think about disease" (Paugh 2014, 227). This chapter draws on her analysis of "the use of ideas about contagion to demarcate racial and sexual difference at sites around the British Empire" (225). Elizabeth Doris Keller summarizes the medical writings on yaws that have contributed to the different perceptions and portrayals of enslaved women, in particular by physicians, planters, and abolitionists. She argues that "by depicting enslaved women as immoral, promiscuous, selfish, and incapable of taking care of their children, patients, and themselves, planters and physicians were able to respectively assert their superiority and necessity as medical practitioners and absolve themselves of blame for the conditions of their slaves by shifting the responsibility for disease and death to enslaved women" (Keller 2022, 3).

Her work illustrates the strongly paternalistic and Eurocentric dimension of colonial medical knowledge through the analysis of scientific publications, correspondence, and elite politics, and sheds light on its usage as a tool of power. Yet many of those who directly dealt with diseases in enslaved people were not necessarily the elite. Professional (i.e., trained) physicians were often not present or available on the smaller islands of the Caribbean, and small plantations may not have been able to afford such practitioners in any case. Medical laymen thus made use of the knowledge of the enslaved or relied on existing expertise (Harrison 2010, 14). Less perhaps through their medical knowledge than their experiences, they often connected diseases to the above-mentioned environmental aspects, leading to different measures of disease control, which similarly contributed to historical processes that have been analyzed under the aegis of the synergies of colonialism and ideas about race.[3]

However, to write on the perspective of the people on the spot, who acted in different roles and often out of self-interest, is not only difficult because of a lack of primary sources; it is equally subject to the reproduction of colonial representations. This is not to mention the lack of sources from the perspective of the people directly affected by colonial atrocities. Any attempt at an empirical reconstruction will always be based on European colonial sources. What is feasible therefore is to highlight the discourses and actions that built fertile ground for racialized patterns of interpretation. Moreover, these perspectives cannot be generalized regarding the highly spatial and social heterogeneity of the different British Antillean Islands. Therefore this paper only features a small part of the aspects of the local heuristic and repertory dimensions of handling diseases, but it does introduce important insights on the Lesser Antilles into the broader debate. Analytically, it relies on the term *environment* to explain the perceived relations between diseases and disease-inducing factors because it best captures the variety of causes that contributed to a particular space being considered unhealthy, including, most importantly, the social component.

To analyze these dimensions, the first section will give a brief overview of the colonial meaning of *reproduction* within the West Indian plantation system during the time of abolitionism in the late eighteenth and early nineteenth centuries and its economic motives. Different measures for containing yaws, some of which were part of more comprehensive strategies of disease control in general, will be highlighted in the second section. It will be shown how economic interpretations of scientific knowledge, which was strongly influenced by the European medicine of the time, had a direct impact on plantation management. Finally, the consequences of these colonial politics and economic interests concerning the perception of enslaved mothers and maternity will be analyzed to point out some of the reverberating effects of colonial approaches to a health problem and its instrumentalization.

ABOLITIONISM, ECONOMY, AND "REPRODUCTION"

Yaws and other treponematoses were treacherous but seemed to not be as deadly as other infectious diseases raging on plantations; people could live with it for years though seldom without symptoms, although these could indicate other diseases or work injuries. This meant that the disease could spread unhindered and often undetected. According to Clement Caines, one enslaver in St. Kitts, the most common nonfatal diseases among enslaved people in the West Indies were ulcers, yaws, and leprosy, with the last causing the greatest damage to health. A proponent of the improvement of living conditions for the enslaved in the colonies—the so-called amelioration[4]—he believed enslaved people suffered from these diseases due to the negligence of duties of care by planters (1801, 150), most of whom delegated medical responsibility to their men on the spot, meaning their administrators or managers. Despite this backdrop of negligence and unnoticed spread, considerable attention was granted to the yaws problem.

One possible reason for this attention was that children were particularly vulnerable: enslaved children were first recorded (in over three-quarters of the cases) as showing the symptoms of the disease when they were between the ages of two and seventeen (Dunn 2014, 80). Moreover, while adults did not necessarily die from the disease, it often seemed to be fatal in children. This posed a major threat to the future of plantation economies. Especially in the period around 1800, when abolitionist movements in the empire were in full swing and it was foreseeable that the (official) slave trade would end, an alternative was frantically sought in order to maintain the labor force on the plantations. The last decades of the eighteenth century were thus marked by a search for strategies to maintain the plantation economy, especially enslaved work. In addition to the bonuses that were given to enslaved women for birthing and caring for their children,[5] an amelioration in the living conditions of mothers and the survival chances of children, and thus a "natural" preservation of labor, became necessary. The fight against mysterious ulcers might not have been as urgent as the fight against diseases that were, de facto, deadly. However, due to the spread of abolitionist views, the focus narrowed on keeping children and newborns healthy so that they could become good workers. The motivation for (medically) securing maternity and childcare was thus primarily economic but also compounded by the wider political situation.

This understanding fits into discussions at that time about the often propagated need of enslaved work for the cultivation of Caribbean resources, revealing the ambivalence behind the medical arguments concerning yaws. While proslavery reasoning claimed that "black" or African people's[6] work—equated with enslaved work—was necessary because their bodies were said to be able to cope better with tropical climates (Brathwaite 1788, 33), investigations into the disease ascertained that it predominantly befell people of African descent. This assertion needed to be explained, and hence a narrative was formed that focused on enslaved women at the center of their environment. For managers' justification strategy of explaining the decline of the labor force and for planters' fight against abolitionist streams, this was a convenient explanation that fit into larger narratives of a "civilizing idea." Adapting persistent arguments of anti-abolitionism, it was suggested that health risks among enslaved mothers could be contained or avoided, for example, by the mothers "attaching themselves to the interest of their family" (Brathwaite 1788, 33), superimposing upon them a normative European notion of marriage and family life.

Moreover, in the context of the alleged race-related susceptibility to disease and the shifting of responsibilities to the mothers, another component was introduced into the broader abolitionist discussion. Observers avoided linking dangerous diseases with white colonial societies, deflecting attention from cruelties against the enslaved, the rampant violent behavior of managers, and the many forms of sexual violence that were associated with the spread of disease on the plantations. Regardless of how diseases were actually transmitted, colonists avoided implying that the institutions of slavery or the slave trade, especially any of the ways through which a plantation was managed, was a possible factor in the spread of a disease. These conflicts of interest were one reason why the spread of the great pox in Europe at the time was often carefully discursively disconnected from the yaws epidemics in the tropics or the problem of venereal diseases in white plantation societies (Paugh 2014, 239).

Against this backdrop, various measures were introduced to contain yaws and other diseases and in this way to increase the number of births on the plantations. As shown, the different perceptions of diseases, diseased persons, and healing methods moved within a discursive framework that was shaped by economic interests. Medical debates were not necessarily scientific, particularly within the plantation setting where disease analysis and control were practice-oriented, but they were influenced by the state of European, that is, Eurocentric medical knowledge of the time. In turn, these different modes of negotiating affected the methods of combating yaws.

DISCOURSES AND MEASURES OF DISEASE CONTROL

Although pathological skin changes occurred constantly on the plantations, the persons in charge of the health of the enslaved often did not know exactly how to manage situations surrounding those conditions. As already mentioned, the diagnosis of yaws appeared often in medical, and more so in plantation correspondences or the like, where it was scrutinized in children and problematized in the context of maternity and childcare. However, it was probably used to describe multiple diseases that caused similar markings on the skin and other identical symptoms, most notably smallpox or leprosy. Contemporaries examined itchy and oozing papules and ulcers, with the former mostly affecting only areas on the lower leg or face. In the second stage of yaws, often weeks later, the soles of the feet, palms, and other parts of the body were afflicted. The later latency stage, followed by destruction of the bones and joints, was not always associated with earlier symptoms on the skin (Dunn 2014, 80).

In order to recognize the forms of contagion, or rather, to prevent them, those responsible made use of a mixture of medical knowledge on the one hand and existing racialized discourses about Black mothers on the other. Most of them emerged from theories that were used to explain why Europeans in the tropics were frequently ill, which was an important factor in connection with, for example, military strategies. "The distinctiveness of warm climates from a medical point of view rested primarily on two observations. First, that putrefaction was accelerated in tropical latitudes by the abundance of heat and moisture. This, and the lush vegetation of the tropics, appeared to be the source of the ubiquitous miasmas which caused Europeans to sicken and often to die. Second, the tropical climate sapped the resistance of newcomers, leaving them vulnerable to such vapours" (Harrison 2010, 64). The medical branch that emerged in reaction to the challenges European soldiers faced in tropical colonies, often referred to as the medicine of hot or warm climates, was the main source of public medical knowledge that was applied in the West Indies. This knowledge was based on the writings of practitioners who believed that forms of adaption to the climate played a decisive role in the development of diseases,[7] and thus it was possible to negotiate the contrariety behind the finding that Africans were allegedly more likely to be affected by a particular disease than Europeans, even though their bodies were said to be more resilient: enslaved people transported from Africa would also have to get used to the environment first, while creole people—those born in the Caribbean—were more resistant. But this idea still did not explain why enslaved children were particularly at risk, when anti-abolitionists tried to veil the inhumanity of the institution of slavery.

These argumentative inconsistencies show the many upheavals of the period and, relatedly, the interplay, but also the complexity, of what was happening in scientific or political discourses. Not least, they mirror how the realities on the estates could play out entirely contrary to perceptions in Britain or by British people in the colonies. Certain islands

Karina Turmann

were thought of as particularly disease prone: the islands of Dominica and St. Lucia were considered as having a harmful climate, whereas St. Vincent was said to have the "best climate" in the West Indies, similar to Gibraltar. Because the interior of St. Vincent is extremely mountainous and well vegetated, many of the sugar plantations lay along the shore, accessible and close to the harbors. Those estates were later described as being cultivated by "miserable looking Negroes" (Dent 1819a). The constellation of sick people working in locations close to the shore, in marshlands, or other wet areas confirmed the miasma theories. Particularly among those on the plantations who lived close to the enslaved, an awareness arose that prophylaxis with a focus on the environment had to become part of the strategy to combat yaws, especially in the lodgings of the enslaved.

One Philip Gibbes, a planter in St. James' Parish in Barbados, claimed that too many of the men in charge on the island did not pay enough attention to the houses of the enslaved, and reproached himself for his own negligence (1788). That the enslaved people should be protected, particularly from moisture, became an urgent concern within the ameliorationist debates: "Light portable sheds to protect [them] from rain are gone too much out of use. I recommend it to you to have them. And I do still more strongly recommend it to you to have proper trays or baskets for the suckling children, with a cover to the heads" (Lascelles et al. 1786, 31). Despite those experiences and the knowledge of the health threats stemming from environmental factors, as well as the neglect of care by plantation managers, the responsibility was shifted to enslaved women. They were accused of inadequate care of their children, of deliberately exposing them to toxic miasmas, and were held responsible when certain measures proved unsuccessful. In the case of yaws control, managers took drastic measures to gain authority over maternal behavior. They locked the mothers up, sometimes with, sometimes without their children, until they recovered. According to Caines, the most effective method was to build a house at a distance from the other slave accommodations where the sick could be treated (1801, 158–59). However, patients were usually given far too little time to fully recover, and this was exacerbated because later symptoms were often not associated with the initial symptoms.

This neglect by management was also reflected in other areas, which is evident in many reports from the colonies; however, proslavery activists often instrumentalized these reports for their own interests. One proslavery polemicist, Gilbert Francklyn, endorsed the theory of physical precondition in the context of adapting to hot climates. He claimed that susceptibility to certain diseases were determined by origin, which amounted to saying that some of the disorders that were widespread in the West Indies were considered to be inherently "African diseases," and that the enslaved people had brought the diseases from Africa to the islands. To the best of his knowledge, none of those diseases was contagious and therefore no creole people would be affected. To explain why no creole people fell ill, but white residents did, he used a polemical logic: Regarding the symptoms that might be observed in whites, he agreed that these diseases were caused by poor and scant diet and by arduous work. He concluded that, while the enslaved were well nourished by healthy food and their work was moderate, the whites were forced to work beyond their abilities without enough and bad-quality food (Francklyn 1789, 54–55).

He could not have given a more contradictory explanation since, at the same time, dietary changes were being found helpful in ameliorating enslaved women's and children's health challenges. Common European remedies like mercury and antimony preparations to treat yaws—all of which were already considered contentious by some physicians—were found to not have any effect, and managers were urged to focus on changing the diet of the enslaved. In contrast to some of the customary practices that, with the kind of physical work that had to be done, did not ensure an adequate supply of food, experienced managers advised, for example, preparing meals twice a day and providing enough food if the enslaved could not provide themselves with food that they grew in their gardens, for example. These approaches seem almost innovative in contrast to the classical European methods of the time but were pervaded by prejudiced ideas concerning the African heritage of the enslaved people. They were criticized for devouring large quantities of fruit and drinking acidic beverages, which were said to cause harmful putrefaction by clogging the stomach and inhibiting both appetite and digestion. Moreover, this acidic diet in combination with treatments with mercury and antimony could, in the eyes of some medical experts, be fatal (see Dent 1819b). In the later formulation of laws for amelioration, the lessons learned from implementing these provisions were considered proof that the strategy had been successful: the change in diet could not prevent diseases, but it would bring about a faster recovery and help develop resistance to them (see "An Abstract of the Meliorating Provisions" 1831, 54). But, again, although those findings were known at the time, colonial interpretations steered away from the scientific evidence and instead embraced the racialized arguments that blamed Black women.

Clement Caines's strategy for treating diseases of the skin reflects this approach, even though he saw himself as an advocate of amelioration who, for example, insisted on reducing workloads. He advised the plantation management that salty food should be avoided as it would hinder healing processes, and to let the affected parts of the body rest by, for example, assigning work that was less physically demanding. Furthermore, the "master himself" should examine the skin every day, and not leave it to the nurses or doctors, ensuring and expanding their control over the enslaved. According to his experience, this strategy would often be better implemented on small plantations, and so they were less affected by yaws (Caines 1801, 159). If the enslaved deliberately "tended" to their ulcers, Caines deemed a few lashes appropriate and necessary, as was common with soldiers when they voluntarily worsened their symptoms in the hospital through drunkenness or the like (153). He brought in these comparisons to make it appear that the enslaved people on his estates were treated no worse than British subjects and addressed the "unnatural sterility" among enslaved women as due to their origin, workload, and mental state, which, he stated, was also responsible for their negligence of their children (267–72).

Caines further advised the managers to always monitor any changes in the body to detect if the patients inflicted wounds on themselves to escape work. If those in charge were better able to read the bodies and minds of the enslaved people, he stated, keeping a close eye on and exerting control over them would be the best measures to prevent bad character traits from taking root (Caines 1801, 155–56). These racialized readings and environmental embedding of social behavior and somatic adaption had a direct influence on the realities of life on the plantations and highlight the strongly paternalistic thinking of the time. Many managers and other colonists were influenced by these theories and shaped the environmental setting on their plantations accordingly, including the enslaved peoples' social space. Moreover, measures used to contain yaws sometimes complicated the social situation on the estate. A manager's decision to send women and children to the yaws building would not only stigmatize their illness, but also suggest bad behavior on their part, which was then used as justification for increased supervision.

As we have seen, the measures of disease detection and control of the spread of the yaws on plantations were part of wider political discussions on both sides of the Atlantic. However, the measures considered and proposed by the proponents of amelioration in Britain could not necessarily be implemented in the colonies, especially on the plantations. Nevertheless, these measures were influenced by the prevailing racialized notions of the need to regulate bodies, climate, and character to contain disease. In the case of yaws, the containment measures were significantly economically motivated. Thus the planters' economic interests, as well as the discourse on disease that was shaped by contemporary knowledge, strongly influenced methods of disease perception and control. While enslaved motherhood was a topic of discussion at all these levels, it was only from a particularly paternalistic perspective. The question remains in which way the already existing prejudices and stereotypes regarding enslaved mothers were reproduced and reinforced, and how that relates to the environmental discourse. The following section attempts to capture this and show that both ameliorative and disease discourses related to the plantation economy were often primarily strategic.

REINFORCING STEREOTYPES

Drawing on their observations and the medical knowledge of the predecessors of European tropical medicine, managers, overseers, and medical staff assumed, or asserted, that yaws inherently befell African or creole people. This theory was underlined by those (absentee) planters who propagated that yaws was a disease of hot climates and could not be spread to Europe via trade networks. The British perception of the disease was increasingly inclined to associate it with a depraved human character. The debate about "reproduction" on the plantations followed this path.

Within the amelioration debates, several causes were found to impede the "natural" increase of the enslaved workforce on West Indian plantations. According to John Brathwaite, an enslaver and agent for the colony of Barbados in London, among them were "promiscuous concubinage; the ambition of making great crops; [and] occasional working the negroes too hard to pay off a pressing demand." Brathwaite claimed that "a manager may sometimes too, unintentionally make the negroes work beyond their strength from a misconception of it" (1788, 29). To encourage the expansion of the workforce, he recommended regulating the annual tax accounting of enslaved people, suggesting "an account of the number of deaths and births should be added and a premium be given in proportion to the increase of the negro stock, and that [Britain] should distinguish by some public mark of approbation, the island in which there shall be the greatest increase by births, in proportion to their numbers" (29).

When asked how many children were born into slavery, he talked of knowing African women who had eight, nine, or ten children, but thought that this was not common. "They begin breeding earlier," he continued, "but do not continue to breed so long as women in [Britain]. I can say nothing of the average number of children they may be supposed to produce" (Brathwaite 1788, 30). On the one hand, his choice of words suggests a strong economic motivation, indicating new regulations by the planters to ensure the maintenance of their workforce. On the other, he discusses the role of Black women and the susceptibility of their children to diseases, particularly in contrast to the whites in the colonies: "The negroes in general are subject to the yaws to which the white inhabitants are not subject, but I cannot from memory say to what other disorder they are subject from which white men are exempt. I cannot say in what proportion negro children are reared" (30).

The allegation that Black women "begin early connections with the sex, and have more variety, which likewise tends to prevent their breeding" (House of Commons 1790, 156) found its way into the House of Commons very early. It stabilized racialized perceptions and discourses related to enslaved mothers in the colonies as well as in Britain, which had a long tradition of portraying Black women in the light of "socio-sexual deviance" (Morgan 2004, 16). The Parliamentary minutes concerning the Antillean island of Grenada also registered that "almost all the children have the yaws, of which, at times, the first medical men in this country have not been able to effect a cure, so as not to leave them liable to other diseases—These are the chief diseases which children are subject to in Grenada more than in colder climates" (House of Commons 1790, 156). The issue was broadly ignored and not a primary concern among physicians or surgeons, who were not actively trying to find a cure for a disease that was not necessarily deadly for "working slaves." In the colonies, to support the claim that the disease was caused by the mothers was convenient and a way for medical and nonmedical personnel to eschew medical responsibility when their prophylactic and therapeutic measures reached their qualitative or quantitative limits.

On the Kittitian Sir Gillies's estate, as on most small plantations, there were no trained European doctors. Generally, "medical treatment for slaves was only extended to working slaves and children who had the potential to grow into productive labourers" (Handler and Jacoby 1993, 75). As on many plantations, enslaved men or women attempted to treat their own medical problems and used healing methods from Africa or the Caribbean as well as others they learned. Although for most illnesses these treatments were more effective than those of European "conventional medicine" and were sometimes adopted by European doctors, in many places they were either devalued or dismissed by the colonial representatives as sorcery or unscientific. With the problem of maternity health, however, not only the enslaved mothers were the focus of attention, but also the enslaved midwives because they were able to draw on a great wealth of medical knowledge. This had a derogatory influence on the colonial perception and treatment of midwives and, once again, women in general. Because cases of pathological skin changes were frequent among children, plantation managers associated their appearance with the presence and practice of midwives, or the mothers themselves.[8] Notably, while both were the target of pejorative stereotyping associated with the illness, promiscuity was the reason found to blame the mothers, while midwives were said to be causing the problem through evil sorcery.

Finally, the discussion around the inoculation of yaws shows another of the many discursive ambivalences on the topic. As already mentioned, suspicion and mistrust toward women heightened and were instrumentalized among the overseers and managers to claim that women would deliberately infect their children. This was a commonly considered hypothesis among colonialists, but one that was less common in European medicine (Paugh 2014, 226). Perhaps Caines was referring to the practice of inoculation in his statement that mothers transmitted the disease to children. The practice likely refers to the general method of inoculating children to prevent serious disease progression in adulthood, which was also used to contain yaws on plantations. The method is known to have been used by enslaved people on the Gold Coast and was also discussed and partially adopted by European doctors. The inoculation of smallpox, for example on slave ships, was already common practice in the eighteenth century. Yaws inoculation among the enslaved was a counterargument to prevailing claims that the disease spread through promiscuous behavior on the part of women. It contributed to the preclusion of "any association of yaws with sexual immorality, causing it to be viewed instead as a childhood disease" (Paugh 2017, 116).

Thus there existed, on the one hand, the narrative of the vulnerability of children, whom women allegedly did not adequately care for, and on the other, the assertion that women's subversive and promiscuous behavior promoted the spread of the disease. Both narratives used the degradation of women to apportion blame and also to scapegoat them when the measures used to curb yaws failed. It is sometimes difficult to reconstruct the extent to which the managers' doubts about the effectiveness of health measures or the accusations of causing bad results in this practice were a targeted form of establishing or strengthening control mechanisms. However, the situation on plantations reveals the utterly paternalistic structures in place for dealing with health issues among the enslaved in general. The depictions of enslaved motherhood and the alleged need for tightened control measures were woven into the recommendations for an amelioration that was portrayed as an improvement of the living conditions of the enslaved.

Overall, to deal with problems related to reproduction of the workforce and the disease the contemporaries referred to as "the yaws," the responsible persons in the West Indian colonies made use of compensatory strategies, due to a lack of knowledge and also because they followed an economic agenda. From the perspective of the colonial agents, while it was clear that disease posed a direct threat to the health of the enslaved, the overriding concern was to ensure their own economic security. This is where the ambivalence between measures and discourses related to disease control became apparent. Despite the high ambiguity of diagnoses, *yaws*, a collective term for pathological skin disorders, appeared comparatively often in plantation correspondences and became an ailment that had to be fought in view of the changing political and economic conditions in the plantation colonies. The contrast between the medical interpretations of yaws and theories used against the abolition of slavery, which suggested that certain groups were less prone to certain diseases and therefore better suited to work on plantations in tropical regions, depicts one of the many ambiguous realms of colonial control. The example of the disease highlights the particular consequences of these discourses regarding the perception of enslaved women in the context of motherhood.

The men-on-the-spot's observations worked hand in hand with established medical theories, such as the miasma theory, to explain where disease originated, locating it firmly in the environment. Despite these findings, the responsibility for diseases and, for example, the high infant mortality rate was transferred to the mothers, underpinned by the argument that especially people of African descent suffered from the disease. The interaction of the different discourses created the "harmful environments" scenario that caused the disease to spread, and one where the mothers' alleged patterns of behavior endangered their own children, exposing them to toxic settings and even intentionally infecting them. By labeling certain environments, in combination with social spaces, as unhealthy, an attempt was made to create a set of strategies for combating certain diseases, but one that endorsed the structural degradation and stigmatization of enslaved women. Despite the prevailing environmental focus at the time, people's heritage and character were the main factors at the core of many of those debates. The assumption that women would use an illness to avoid work reinforced these already existing stereotypes. The fact that a disease that at first seemed rather less threatening in the Antillean colonies, then took on special significance once concerns over economic viability on the small islands and plantations arose, illustrates a moment in which a shift in the imagined relationship between environment and disease becomes visible, one that Pratik Chakrabarti describes as a shift from "climatic theories of disease" to "climatic determinism of disease and race" (2014, 65–66).

It was shown that the paradoxical dynamic of the regulatory attempts to implement economic interests within abolitionist interests and to remodel plantation environments was neither extraordinary nor in any way a positive effect of ameliorationist ambitions. This form of colonial management was rather used to gain control over various parts of women's lives and promoted the ongoing dehumanizing portrayal of the problem of yaws that deprived women of their individuality and emotions. Their experience of maternity and motherhood was marked by pain, grief, and everyday loss (Morgan 2004, 108).[9] The claim that mothers deliberately infected their children and put them in danger not only downplayed the overall experience of violence of enslaved people, but also negated the influence and adoption of African knowledge and practices by European medicine.

The threat that was posed by certain environments on plantations in the West Indies thus consisted of a reciprocal relation between discourses and measures of disease control. Since they were operating in a colonial context, these did not manifest themselves as a governmental strategy in the modern sense but were strongly conditioned by local circumstances where the situatedness of the health problem was reflected. While they were never entirely free of political leverage, especially in times of abolition, the implementation of measures on the spot was nevertheless mostly subject to situational circumstances and driven by plantation officials. Through attributing the susceptibility to, and the spread of, diseases to certain degrading physical and mental human characteristics, establishing those dangerous environments influenced the perception of and dealing with already stigmatized groups. Not least, this example shows how highly racialized the environmental debates around health and disease have been, which has long shaped the historical thinking on disease and environment and in medical historiography (see also Arnold 1998; Nash 2017). In the case of yaws, this was particularly evident in the case of enslaved mothers or women in general whose situation could well be made worse by colonial health control measures. The identification of symptoms was quite differentiated, and it sometimes seemed as if action was taken on the basis of certain environmental circumstances. However, when considering the example of the spread of the disease among children, while it was not wrong in its mere medical analysis, we can see that the attempts at explanation and containment measures were constantly adapted to degrading interpretations. Source criticism shows us that scientific or medical interpretation and political or economic interests were mutually conditioned, and how the colonial, in the form of power, violence, and exclusion, was a decisive factor in the creation, interpretation, and control of epidemic diseases.

CHAPTER 3

"A DISEASE OF PLACE"

THE ECOLOGICAL UNDERPINNINGS OF US PELLAGRA RESEARCH, 1902-1914

Julia Engelschalt

In recent years scholars in medical as well as environmental history have repeatedly called for an integration of their two fields of inquiry. In her widely acclaimed 2006 book, *Inescapable Ecologies*, environmental historian Linda Nash provides a compelling rationale for this convergence. Despite the best efforts of self-proclaimed modern biomedicine to pretend otherwise, she writes, human bodies have never been independent from their environments. By consequence, "we are forced to acknowledge that humans are not simply agents of environmental change but also objects of that change. Conversely, the environment is more than an object upon which change is enacted; it is also an agent of sorts that acts upon the bodies inhabiting it" (Nash 2006, 8). In 2018 the works of Nash and others were taken up in a programmatic article by medical historian Christopher Sellers, who identifies a tendency within twentieth-century medicine toward what he calls "place neutrality," that is, the notion that disease can (and must) be diagnosed and treated within the human body, without reference to its surroundings. Instead of co-opting and thereby reproducing the paradigm of place neutrality, Sellers urges both historians of medicine and the environment to "understand the past two centuries of medical history in terms of a seesaw dialogue over the ways and means by which physicians and other health professionals did, and also did not, consider the influence of place—airs and waters included—on disease" (Sellers 2018, 6).

The debate over the etiology and epidemiology of pellagra in the United States during the first third of the twentieth century uniquely lends itself to discerning precisely this "seesaw dialogue" between place-neutral, pathogen-oriented theories on the one hand, and more ecologically oriented spatial ontologies of disease on the other. Pellagra—known today as a dietary deficiency disease caused by a lack of niacin, or vitamin B3—was brought to the attention of physicians and public health officials in the early 1900s as a debilitating affliction which particularly affected the rural poor. Although it quickly spread across the entire country, it was most prevalent in the southern states, thus contributing to the long-standing narrative of the "diseased South" (Savitt and Young 1988; Ring 2012). Since its first occurrence in southern Europe and France in the eighteenth century, medical experts had been searching for its cause. In the late nineteenth century it was commonly held that the excessively corn-based diet of poor farm laborers was somehow implicated in its etiology. By the time it was recorded in the United States, however, to some experts the recent rise of the germ theory of disease seemed to suggest the presence a microbial agent (Bryan and Mull 2015, 26). Much of the historiography on pellagra focuses on the controversy between so-called zeism (from *Zea mays*, the scientific name for corn) and those multiple other theories that downplayed or fully denied the role of corn in the etiology of pellagra—a controversy that would eventually be settled by Joseph Goldberger, a Public Health Service (PHS) researcher who was able to establish the nutritional origin of the disease around 1915 before niacin would eventually be discovered in 1937 (Etheridge 1972; Roe 1973; Gentilcore 2015). Representative of "place neutrality" in medicine, the germ theory and its emphasis on microbial causative agents is, in those accounts, regarded as a hindrance to the discovery of the true cause of pellagra (Leslie 2002; Bryan and Mull 2015).[1]

This chapter seeks to complicate this well-rehearsed narrative of pellagra in the American South by taking a closer look at the multiple ecological considerations that pervaded debates over its etiology and epidemiology as well as their practical consequences. Following a brief overview of European pellagra etiologies prior to its emergence in the United States, my main focus lies in the influence of those theories, particularly Louis Westenra Sambon's notion that pellagra was an insect-borne disease, on American pellagrology until about 1914. The great interest taken by early twentieth-century American medical experts in Sambon's theory has commonly been attributed to the fact that it resonated with a growing belief in the germ theory and in disease specificity, especially because its epidemiology seemed to suggest that pellagra, like tuberculosis or bubonic plague—whose causative microbial agents had recently been discovered—was an infectious disease (Gentilcore 2015, 34–35). What this explanation does not take into account, however, is the strong appeal of Sambon, not only as a proponent of modern, microbially oriented laboratory medicine, but as an expert in *tropical* medicine. This distinctly place-oriented subdiscipline—which did not reject, but readily incorporated the insights of laboratory bacteriology—had become an extremely relevant field in the United States since the acquisition of tropical overseas territories in the aftermath of the Spanish-American War of 1898.

Throughout their (unsuccessful) attempts to find a microbial agent for pellagra and potentially confirm Sambon's hypothesis, American medical experts simultaneously understood pellagra as "a disease of place," as PHS passed assistant surgeon Claude H. Lavinder remarked in 1911 (Lavinder 1911, 1462). In doing so, they perpetuated a long-standing discourse revolving around the construction of a (semi)tropical disease ecology of the American South shaped by multiple environmental factors such as vegetation, water supply, agriculture, and climate. This understanding of place was highly selective and somewhat arbitrary in its emphasis on those seemingly "natural," unchanging environmental conditions. The "human factor"—the situatedness of humans and their quotidian health-related, nutritional, and agricultural practices within these environmental conditions—tended to play a subordinate role in these disease ecology considerations and was more present in the works of researchers who were interested in finding a microbial explanation for pellagra. Nonetheless, I will demonstrate that the boundaries between humans and their environments, between etiology and epidemiology, and between microbial and ecological approaches to pellagra were exceedingly difficult to maintain. Rather than pitting ecological and microbial concepts of disease causation, contagion, and control against each other—with the latter typically being portrayed as a more progressive, modern, and rational perspective—both should be regarded as mutually intertwined strands of thought that complemented and reinforced each other in the context of early twentieth-century American pellagra research.

PELLAGRA ETIOLOGIES IN EUROPE BEFORE 1900

By the time Atlanta-based physician Henry F. Harris reported the first officially diagnosed case of pellagra in the United States in the spring of 1902, the disease had been known in Europe for about 150 years (Harris 1902). Its symptoms were first described in 1735 by the Spanish physician Gaspar Casal during a period of economic downturn and mass hunger in the Iberian peninsula (Rajakumar 2000). Casal's patients were impoverished peasants who displayed what was later to become known as the four Ds of pellagra: dermatitis, dementia, diarrhea, and, in many cases, death.[2] Over the course of the eighteenth and nineteenth centuries, the affliction was observed across southern Europe, including the French Midi, parts of Italy, the Austrian region of Carinthia, Hungary, and Romania (Roe 1973, 28). Especially in Italy, physicians quickly became suspicious of maize, a relatively recent transplant from the Americas that gradually replaced other staple crops and soon constituted the better part of the diet consumed by poor rural populations (Gentilcore 2013). However, governments were often reluctant to acknowledge the existence of large-scale poverty under their auspices and therefore tended to support theories that implicated causes other than the concomitant problem of mass malnutrition (Roe 1973, 60).

Beginning in the late eighteenth century, Italian pellagrologists like Francesco Fanzago and Giambattista Marzari discussed the influence of climate, environment, and even germs—albeit in a pre-Pasteurian understanding—on the etiology of pellagra (Roe 1973, 55–57). Nonetheless, their works laid the foundation for the so-called zeist theory: despite various theoretical differences, they agreed that a heavily corn-based diet was involved in one way or another (Gentilcore 2013, 65). Soon some physicians in various countries contended that the culprit was not domestic corn, but rather spoiled corn imported from abroad—a protectionist argument that would resurface in the American context several decades later (Roe 1973, 61). Cesare Lombroso, an army surgeon, phrenologist, and criminologist who became known for his infamous writings on the alleged physiognomy and hereditary character of criminal behavior, initially embraced hereditary theories of pellagra—a less prominent, yet quite persistent strand of theorizing that emerged in the mid-1850s and would be taken up in the writings of early twentieth-century American eugenicists like Charles B. Davenport and Elizabeth Muncey (Roe 1973, 51, 64).

In the mid-nineteenth century new insights from the laboratories of Louis Pasteur, Robert Koch, and others ushered in so-called germ theories of disease, that is, a set of assumptions revolving around the notion that each infectious disease is caused by one specific microorganism (Worboys 2007). The rise of the bacterial theory of infection was quickly taken up in pellagrology and most prominently espoused, once more, by Lombroso. The spoiled-corn hypothesis, also known as toxicozeism, held that it was not maize per se, but a yet unknown pathogen—potentially a mold or fungus—that induced pellagra in those relying on a corn-based diet (Gentilcore 2015, 24). The toxicozeist theory "was derided at first, but soon came to dominate medical discussions" because it "had laboratory science on its side," and because "it suited government, which sought a clearly identifiable enemy and practicable solutions" (Gentilcore 2015, 24). Very early on debates over the etiology of pellagra were thus shaped by European governments' perception of the disease as a threat to labor productivity and food security. At the same time state institutions favored an etiological explanation that suggested the future possibility of controlling its yet unknown causative agent by technocratic means. Indeed, in 1902 the Italian government issued a new law based on the zeist theory that prohibited the distribution of spoiled corn, "led to the setting up of soup kitchens, health stations, and pellagra hospitals," and resulted in an overall decline of pellagra incidence in Italy during the following years (Gentilcore 2015, 25).

Despite the apparent practical applicability of Lombroso's theory, some medical experts continued to search for other possible causes of pellagra. One of the most influential proponents of anti-zeism, who would also become relevant in the American context, was Louis Sambon, a researcher in tropical medicine who had previously investigated vector-borne infectious diseases, who maintained close connections to the British malaria researcher Ronald Ross, and quickly gained a reputation at the newly inaugurated London School of Tropical Medicine in the early 1900s (Gentilcore 2015, 27–29). Following two extended visits to several Italian regions with a high incidence of pellagra in 1900 and 1903, Sambon denied the Lombrosian doctrine of infected maize and instead developed a theory implicating the *Simulium* fly (sand fly) as the vector of a putative pellagra pathogen. This claim was based on two observations: the apparent seasonality of pellagra, which Sambon attributed to the life cycle of the insect in question, and the alleged congruence of the preferred breeding grounds of sand flies—areas near small bodies of water—with areas of high pellagra incidence (Roe 1973, 86–87). According to Gentilcore, both factors "made pellagra similar to malaria, sleeping sickness, Rocky Mountain fever," and other insect-transmitted diseases Sambon had previously encountered (Gentilcore 2015, 31). Although Sambon notoriously failed to empirically substantiate his claims and was ridiculed by many of his Italian colleagues, who had witnessed the successful application of zeism, his vector theory was well received at the 1905 annual meeting of the British Medical Association and soon became one of the most cited works on pellagra on the other side of the Atlantic (Roe 1973, 86; Sambon 1905).

PELLAGRA RESEARCH AND AMERICAN EMPIRE, 1902-1914

While Sambon was developing his vector theory of pellagra, the rapid spread of the disease in the United States had begun to raise concerns among physicians and state public health officials, particularly in the southern states. In 1907 George H. Searcy, a physician at the Mount Vernon Insane Hospital in Alabama, claimed to have treated "some three or four cases of this disease" every summer for several years. Like Harris in 1902 Searcy believed that pellagra was caused by the consumption of "damaged corn" along with general malnutrition and bad sanitary conditions resulting from extreme poverty (Searcy 1907, 37). The prognosis was "always unfavorable," with "death ensuing in most cases within from two to three weeks" (38). With more and more states reporting a rapid increase of pellagra incidence around the same time, physicians and public health experts eagerly began to investigate its etiology, as is evidenced by the fast-growing amount of research articles, reports, and investigative committees on the subject. A first conference was held in December 1908 in South Carolina, followed by three triennial nationwide conferences organized by the National Association for the Study of Pellagra in 1909, 1912, and 1915. In 1909 the PHS set up laboratories at the South Carolina Hospital for the Insane in Columbia to investigate pellagra under the leadership of Claude Lavinder. Meanwhile, private philanthropy had also taken an interest in the disease, resulting in the founding of the Illinois Pellagra Committee (1909–1911) and the Thompson-McFadden Pellagra Commission (1912–1917), which collaborated with the New York Post-Graduate Medical School in field research projects in Spartanburg, South Carolina, and in the British Caribbean (Bryan and Mull 2015, 30).

Given the spread of pellagra among people living in close proximity to one another, as was the case in Searcy's institutionalized patients, and because "late nineteenth-century enthusiasm fueled a chase for infectious etiologies," it seemed almost inevitable for medical experts to believe that they were confronted with an infectious disease caused by a microbial agent (Bryan and Mull 2015, 26). However, in contrast to their Italian colleagues, American researchers and physicians were less inclined to subscribe to the toxicozeist approach, despite the fact that it neatly aligned with the microbial paradigm. Two leading figures in the American pellagra debate, Claude H. Lavinder of the PHS and James W.

Babcock, superintendent of the South Carolina State Hospital for the Insane, translated and edited Armand Marie's abridged French version of Cesare Lombroso's 1892 *Trattato profilattico e clinico della pellagra* in 1910 in such a way that it included "the latest opinions regarding the possible parasitic origin of pellagra" (Marie 1910, 3–4). Like his PHS colleague Joseph F. Siler, Lavinder had previously visited Italy, where he had come into contact with Louis Sambon and his vector theory of pellagra (Roe 1973, 87).

Sambon's influence on American pellagrology was by no means a coincidence, but points directly to the imperial dimension of pellagra research. In contrast to Lombroso, Sambon was a renowned (albeit controversial) expert in tropical medicine, a discipline that had only recently been established after the United States had acquired several tropical overseas territories—among them Cuba, Puerto Rico, the Philippines, and (as of 1903) the Panama Canal region—following its victory in the Spanish-American War of 1898. Much like their British, French, German, Dutch, and Japanese colleagues, American medical and public health experts became increasingly interested in tropical diseases, both for the purpose of protecting white American troops and civilians in the colonial project and in the spirit of humanitarian uplift and "benevolent assimilation" vis-à-vis their newly colonized subjects (Anderson 2006; Schumacher 2016).[3] Lavinder, Siler, the future surgeon general Rupert Blue, and many other high-ranking medical experts had either served as army surgeons during the war or otherwise made a career in the context of the budding American branch of tropical medicine after the war of 1898 (Anderson 2006, 229–31). The discipline was much less firmly institutionalized than in Great Britain and other European colonizing nations; therefore many former military medical officers worked for the PHS, which grew directly out of the US military, or took their expertise in tropical medicine into other related fields (Farley 1991, 31–33).

In addition to those continuities in personnel, Sambon's theory resonated in many ways with the actual practices of disease control that pervaded tropical public health in the early years of American overseas colonialism. At the turn of the century, a group of army physicians around Walter Reed was sent to Cuba in order to determine the exact cause and transmission of yellow fever and was able to confirm the theory of Carlos Finlay, a Cuban physician who had proposed that the disease was insect-borne (Espinosa 2009). As a consequence, the Army Health Department under Colonel William C. Gorgas began to conduct a systematic antimosquito campaign, establishing practices that were subsequently applied during the construction of the Panama Canal, assisted by a host of entomologists who traveled to the region and conducted surveys of the local insect fauna (Sutter 2007). Another pivotal impulse came from Puerto Rico, where US Army surgeon Bailey K. Ashford identified *Ancylostoma duodenale*, also known as parasitic hookworm, as one of the main causes of what had hitherto been known as "tropical anemia" (Ettling 1981, 29). With the help of Gorgas and Ashford's former professor, Charles W. Stiles, these insights were relayed back to the US mainland, where they helped to dismantle the long-standing narrative of a "germ of laziness" that was believed to infect poor rural southerners and was often used as an explanation of alleged southern backwardness (Immerwahr 2019, 137–43). To many public health officials, the success of the southern anti-hookworm campaign conducted by the PHS from 1909 on constituted an exemplary lesson in building trust in medicine among poorly educated, hookworm-afflicted rural populations (Etheridge 1972, 14–15).

As medical experts grappled with very different types of disease against the backdrop of American imperialism, one common conceptual denominator for all these campaigns was the notion that disease—whether microbial, parasitic, infectious, or not—ultimately had to be understood and treated within a place-based paradigm of endemicity, taking into account any number of strictly locally oriented variables: soil conditions; water supply (including human-made sanitary infrastructure); altitude; and various climatic factors such as sun radiation, precipitation, and the like. By correlating insect breeding grounds with pellagra incidence, researchers like Sambon and his American followers were thus speaking directly to some of the most basic tenets of tropical medicine. The fact that the vector hypothesis could not be substantiated did not pose an obstacle to scientific research. In fact, it even contributed to its immense popularity precisely because it forced the pellagrological community to continue searching for supporting evidence. By designing and adjusting their survey techniques, considering different insect species as possible vectors, and including a variety of environmental factors in their investigations, pellagra researchers thus produced an ever-growing amount of data in the attempt to eventually deliver proof for the vector hypothesis.

Distinctly microbially oriented hypotheses about the etiology of pellagra—zeist or other—were thus not necessarily regarded as a contradiction to the practice of examining the specific localities and environmental circumstances in which people were afflicted by pellagra. On the contrary, microbially and environmentally oriented considerations tended to overlap and complement each other in etiological and epidemiological research. For instance, Eugene Bondurant, a professor of nervous and mental diseases at the University of Alabama in Mobile, claimed

Julia Engelschalt

in 1909 that what he called a "toxemic disease" was "favored by insufficient and unsuitable food, unhygienic surroundings, and a warm and moist climate" (Bondurant 1909, 300). In similar fashion, R. M. Grimm, passed assistant surgeon with the PHS, reported on pellagra in Georgia, where he found the climate to vary "from temperate and never extreme in the mountains to subtropical in the southern portion" (Grimm 1913, 443). A particularly telling example is Lavinder, who claimed in 1911 that, regardless of where one stood on the issue of zeism, "evidence is accumulating that the disease is one of locality or place" (Lavinder 1911, 1462). While Lavinder has been regarded as a supporter of zeism, several letters he wrote to Rupert Blue in 1912, shortly after the latter's appointment as surgeon general, suggest that Lavinder was deeply impressed by Sambon's vector hypothesis (Gentilcore 2015, 37). In late January Lavinder noted that epidemiological research on pellagra had to include entomological groundwork in affected regions (Lavinder 1912a). A few days later he emphasized his claim by pointing out the obvious seasonality of the disease, thus echoing Sambon's earlier observations. In order to investigate "Sambon's important hypothesis," he wrote, "studies must of necessity include a survey of the biting insects common to the places in which pellagra is seen." Generally such entomological studies were already underway, but Lavinder insisted that "the studies of cases and entomological studies should go hand in hand, and not independently as now" (Lavinder 1912b).

Indeed, a year earlier Kentucky state entomologist Harrison Garman had studied the correlation between pellagra incidence and the presence of various flies of the *Simulium* genus in his home state. However, his 1912 report explicitly stated that "none of the species found by Dr. Sambon in regions in Italy in which the disease is prevalent occurs in the United States." Garman made it very clear that he had "not at any time committed himself to the 'Insect theory'" and was "not satisfied with the evidence" he had gathered in the course of his research (Garman 1912, 3). Unconvinced that absence of proof also meant proof of absence, Garman continued his work by considering various other insect species as potential carriers of pellagra. In his capacity as a botanist, moreover, he minutely examined several locally collected ears of corn and commented on the unhealthful properties of those he found to be spoiled. His report also contains ample illustrated material, including drawings of various insect species as well as photographs of corn husks, families afflicted by pellagra, and locations where he had collected specimens. Though certainly far from empirically robust in terms of its epidemiological approach—after all, the human dimension of pellagra was not Garman's primary field of expertise—it appears that early studies on the etiology of pellagra revolved around its ecological dimensions precisely *because* researchers were looking for a pathogenic cause and a vector that they hoped to contain and potentially eradicate on location. Echoing the technocratic fervor of Progressive Era public health reform, this approach to disease control presaged the eradicationist paradigm that would come to characterize national as well as international public health work in the context of Rockefeller's medical philanthropy and the early World Health Organization (Cueto 1995; Elman, McGuire, and Wittman 2014; Birn 2014).

Shortly after the publication of Garman's report, Claude Lavinder of the PHS informed the Bureau of Entomology at the US Department of Agriculture about the situation in the counties of Spartanburg, Chester, and Rock Hill, which he claimed were "marked pellagra centers" in the state of South Carolina (Lavinder 1912c). Between June and October 1912 the bureau sent two of its employees, Allan Jennings and W. V. King, to South Carolina to collaborate with members of the Thompson-McFadden Commission in an "investigation of the premises of pellagrins and the neighborhoods in which they resided, with special reference to the presence, distribution, and biologies of such insect groups and species as appeared worthy of consideration" (Jennings and King 1913, 411). They took note of the "sanitary and other conditions which might have a bearing upon the presence or abundance of insects," interviewed afflicted individuals and their households, and inspected "the more remote surroundings" to the dwellings of known pellagra patients with particular regard to the presence of swamps and marshes (whose influence they found to be negligible) (412). Based on their knowledge of vector-transmitted diseases, they proposed two main arguments for the vector hypothesis in pellagra: the appearance of "isolated sporadic cases" with no prior contact to other affected patients, and the fact that the disease seemed to spread among family members following a "definite chronological connection" (415). After a lengthy consideration of various insect species, including mosquitoes, fleas, and *Cimex lectularius*—better known as bed bugs—Jennings and King tentatively concluded that the only possible insect vector matching the temporality and spatial distribution of pellagra was the stable fly (*Stomoxys calcitrans*) (439–40).

The tropical medicine–inspired search for a pellagra vector also led researchers beyond the contiguous territory of the United States and into the British island colonies in the Caribbean. At Sambon's invitation, Allan Jennings and Joseph Siler conducted an investigative mission in Barbados under the auspices of the Thompson-McFadden Commission in September 1913. Siler's 1915 report of the mission echoed a long tradition of medical treatises on diseases in "hot climates" by first giving a detailed account of the climatic and geographical conditions on the island before elaborating on pellagra incidence among the local population, which had significantly increased over the past decade, and on the entomological findings obtained by the author and his colleagues (Siler 1916). The document epitomizes the fervor with which (especially privately funded) American pellagrologists sought an etiological explanation that downplayed the crucial role of poverty and chronic malnutrition in the rise of pellagra that had occurred over the preceding decade (Bryan and Mull 2015, 30–31). The fact that Siler and his contemporaries were not inclined to attribute more than a predisposing role to malnutrition in the etiology of pellagra may partly be explained by the contemporary state of nutritional research, which had only recently begun to engage with concepts such as nutrient groups and vitamins (Thompson 1902; Haushofer 2022). However, by neglecting the dietary dimension of pellagra in favor of a putative microbial causation, members of the Thompson-McFadden Commission were also following the political and economic interests of the agricultural industry.

Siler and his colleagues extensively reviewed reports and statistics produced by local medical authorities, examined urban as well as rural settlements with regard to their dietary and sanitary conditions (which they found to be better than expected, but still deplorable), and collected insect specimens across the island. In his discussion of the delegation's findings, quite unsurprisingly, Siler concluded that corn could not be the culprit; after all, he claimed, the inhabitants of the island were "now eating the same variety and quality of staple foodstuffs as were used by former generations" (Siler 1916, 221). To Siler the fact that most of the corn consumed in Barbados over the previous decade had been imported from the United States did not deserve further consideration because the quality of American corn—his implicit assumption—was beyond reproach. Then again, entomological research did not yield any connection between pellagra and the sand fly, which "does not exist in Barbados," thereby discarding Sambon's hypothesis that pellagra was caused by protozoa and transmitted by an insect vector. The only possible explanation then was what the Thompson-McFadden Commission had believed all along, but remained unable to prove empirically "that pellagra is an infectious disease communicated from person to person, possibly through contamination of food with the excretions of pellagrins" (221).

Despite repeated and increasingly desperate attempts to make its etiology align with the microbial paradigm, the notion of pellagra as a disease of place did not vanish from sight. As many of the above examples indicate, ecological thinking pervaded American pellagra research, particularly during its first decade. The term *ecology*, however, is largely absent in contemporary medical literature. One exception can be found in a 1914 study of pellagra by Stewart R. Roberts, an Alabama physician and former professor of biology at Emory College. In a section discussing the potentially infectious nature of pellagra, Roberts compiled a list of "ecological evidence" to complement his previous considerations of pathological findings. The first two points speak directly to the putative impact of climate on the etiology of pellagra that, he claims, "occurs in tropical and subtropical climates where infective diseases, and especially diseases caused by parasitic protozoa and parasitic worms, are prevalent. Wherever pellagra occurs, malaria and hookworm disease are to be found, and nearly always amebic dysentery" (Roberts 1914, 250). Moreover, the author marks pellagra as "a rural disease" and, following Lavinder and Sambon, as "a disease of place" (250, 252). The seasonality of the disease was frequently put forward to argue that pellagra was infectious. Like many of his colleagues, Roberts supported this claim by referring to the unequal gender distribution of pellagra (with women being more affected than men) and to the fact that infants—who did not yet eat a full diet and thus could not have contracted the disease through spoiled food—could also display the disease (250–51). Such generalizing statements on the basis of contemporary disease statistics should be treated with caution; after all, the availability of public health infrastructures, access to those infrastructure, and the availability of statistical data on health and disease were anything but reliable in the early twentieth-century American South (Marks 2003; Byrd and Clayton 2000). As Roberts's account shows, ecological thinking was not necessarily regarded as a contradiction to the infectious paradigm, but rather as valid evidence potentially supporting the vector hypothesis and other microbially oriented etiological theories.

The microbial paradigm certainly dominated mainstream early twentieth-century medical theorizing and practice in the United States. However, the field reports and articles examined here indicate that the research practices applied in the process of solving the mystery of pellagra's etiology frequently transgressed the boundaries between human bodies, putative disease vectors, and their shared environments, establishing a fluid disease ecology in which natural, social, and nonhuman environments overlapped and intersected in multiple ways (see Jones 2004). In that sense researchers like Roberts, Garman, and various members of the PHS may be said to have foreshadowed the network of disease ecologists that, as Warwick Anderson has demonstrated, would emerge later in the century in opposition to the purely mechanistic, human-centered approach of the "microbe hunters" (Anderson 2004, 39; Honigsbaum 2017). Though certainly not deliberately holistic in their approach to pellagra, the investigators presented here did embrace an ecological perspective of the disease that was steeped in the environmentalist tenets of tropical medicine in the context of American imperialism (Anderson 2006; Reyes 2014). It was particularly in this context that tropical environments and their indigenous inhabitants were subsumed under a conceptual imaginary of disease residing in specific localities and, by extension, the people populating them. Though certainly less overt in the domestic context of pellagra research—after all, southerners were citizens of the United States and could not explicitly be treated as colonial subjects—we can nonetheless observe a similar spatial paradigm through which the American South and its impoverished rural population became part of one and the same disease ecology framework.

In the following years the history of pellagra etiology would become inextricably linked with the name of Joseph Goldberger, who was sent to the American South by the PHS in 1914 to further investigate the cause of pellagra. Goldberger's peculiar epidemiological approach, which combined socio-ecological and nutritional research, was in many ways an expression of the growing pressure of the "social question" that manifested itself during the late Progressive Era (Robertson 2015, 45–47). While his experiments certainly corrected numerous methodological shortcomings of earlier studies by relying on controlled environments in which Goldberger was eventually able to induce pellagra in individuals by diet alone (while attempts to "infect" patients with the disease famously failed), the notion of pellagra as a "disease of place" lingered for years (Rajakumar 2000, 275; Flannery 2016). As late as 1922, in a report submitted to Surgeon General Hugh S. Cumming, Goldberger noted that for several weeks he had been "occupied chiefly with the examination of data relating to the climate of certain Old World localities in which pellagra has long been highly endemic." The reason for this was, he explained, "that the climate of such localities as Roumania [*sic*] and Bukovina corresponds fairly closely to the climate of Massachusetts for example. This is of interest in its bearing on the question of the relation of climate to pellagra prevalence" (Goldberger 1922).

This chapter has shed light on the ecological underpinnings of early pellagra research in the United States during the "pre-Goldberger" era. While the term *ecology* itself was rarely used in this particular discursive context, the practice of pellagra research very much revolved around multiple aspects of the relationship between individual pellagra patients, their communities, and their nonhuman environments. Conceived as a disease of place, the etiology of pellagra was investigated with regard to climatic and geographical conditions, to its nutritional dimensions (even if only to disprove the zeists), and to its potential origins in the insect fauna of the localities in which it appeared. This did not, as one might assume, happen in contradiction to microbial- or parasitic-oriented research, but was fully in line with the growing appeal of laboratory methods and the microbial paradigm of disease causation. Indeed, as the popularity of Sambon's *Simulium* hypothesis among American medical experts shows, particularly under the impression of recent successes in tropical medicine, researchers were more than ready to embrace the notion of a specific causative agent—a typical element of what Christopher Sellers (2018) calls "place neutrality" in medicine—while simultaneously investigating the climatic and environmental conditions that they believed to be relevant for the high prevalence of pellagra in the American South.

To be sure, the ecological framework of early pellagra research was not only motivated by an interest among members of the medical and public health community in the relationship between health, disease, and the environment. The overwhelming prevalence of pellagra in the American South strongly and immediately resonated with actors who had vested political and economic interests revolving around the health of the rural workforce as a crucial factor in agricultural production for domestic and foreign markets. At the same time national and state authorities—especially in the southern states—were often unwilling to acknowledge the problem of mass poverty and food insecurity among rural laborers. Therefore they tended to promote and fund pellagra research focusing on a potential microbial or parasitic origin of pellagra. The dietary branch of pellagra research eventually pointed to a failure of national and state governments to guarantee food security and general livelihood to a large part of their populations. Even though Goldberger and his colleagues substantiated their dietary hypothesis with irrefutable experimental evidence, their findings would only gradually be accepted by political and economic actors because they suggested that the successful control of pellagra could be achieved only through a concerted effort to improve the standard of living and food supply in the American South.

CHAPTER 4

ECOLOGIES OF VIOLENCE IN URBAN PERIPHERIES DURING HEALTH EMERGENCIES

Oswaldo Santos Baquero, Sara Cristina Aparecida da Silva, and Júlia Amorim Faria

The COVID-19 pandemic has strained the lives of those residing in urban peripheries, but not only due to the pathological effects of the virus. The pandemic has rather intensified a network of violences that had already affected humans and other animals, as well as environments indispensable for good living (*buen vivir*). The urban and public security policies that produce and maintain urban peripheries are part of a historical and structural process of this network. They impose precariousness, exploitation, persecution, territorial expulsions, incarceration, and domestic aggressions that harm peripheral subjects—human and other-than-human. Health emergencies, such as COVID-19, further intensify this dynamic and have given rise to a syndemic of violences in urban peripheries. In syndemics two or more epidemics occur simultaneously and synergistically. Therefore in a syndemic of violence at least two forms of violence co-occur at an epidemic magnitude and potentiate each other.

Several reasons, including violence, can lead to the declaration of a health emergency. However, emerging infectious diseases are more often the reason health emergencies are declared. Such diseases are characterized by being either new, preexisting but increasing in incidence and/or geographic spread, or caused by pathogens categorized as a priority (Mackey et al. 2014). The possibility of the emergence of a highly transmissible infectious disease with pandemic potential was expected even before COVID-19, even though it was not known exactly which etiological agent would cause it, when it would occur, or how serious it would be (Gates 2018). It was also known that most emergent diseases are zoonotic (a spillover from other-than-human to human animals). Similarly, new health emergencies are expected to be triggered by zoonoses in the future and their impact aggravated by preexisting violence. It is also expected that the consequences and resources mobilized to confront them will, once again, be unevenly distributed across multispecies collectives.

Syndemic health emergencies are not new. The syndemic of epidemic infections and violence is historically frequent. Ebola is a recent example of how dealing with a virus, a disease, or an infectious emergency can be influenced by historical, colonial, and violent processes that continue to shape health emergencies. As many have pointed out, unattended basic needs, weak health systems, civil conflict, and intersectional marginalization ensue from the purposeful underdevelopment imposed on colonized countries, where health "aids" also conceal unethical research and industrial interests, increasing local distrust of international interventions, which deepened, for example, the Ebola burden (O'Brien and Tolosa 2016; Sirleaf 2018; Cohn and Kutalek 2016; Nguyen 2019; Wilkinson and Leach 2015; Bardosh, Leach, and Wilkinson 2016; Richardson 2020). Prevailing discourses for dealing with epidemics insists on health emergency preparedness through timely detection of pathogens and the mitigation of their spread, while systematically and ideologically omitting structural causes for the emergence of these pathogens and disregarding inequalities in the capacity various actors have to adopt preventive, protective, and recovery measures (Richardson 2020; Farmer 2004; Wallace et al. 2020).

The burden of violence and pandemics such as COVID-19 is higher in urban peripheries (Raposo et al. 2019; Araujo 2001; Bermudi et al. 2021), and that burden is exacerbated by its demographic context. According to the United Nations, in 2018, one in four humans of the world's urban population lived in a favela, totaling more than 1 billion persons (United Nations 2021). In Brazil between 2010 and 2019 the number of favelas increased by 108 percent (Instituto Brasileiro de Geografia e Estatística 2020). In 2019 more than half of the households in capital cities, such as Belém and Amazonas, were located in favelas (55.5% and 53.4%, respectively), while there were smaller cities such as Vitória do Jari that registered even higher numbers, with three out of four households (74%) (Instituto Brasileiro de Geografia e Estatística 2020). Also in 2019 in the city of São Paulo 12.9 percent of households were in favelas, a significant percentage of its 12.25 million inhabitants. It should be noted that these data do not include the favelas that emerged as a result of the COVID-19 health emergency.

Unfortunately, there are almost no data available regarding other-than-human populations exposed to violence in urban peripheries. In our research to help close this gap, we found in a recent census that in two groups of favelas in Brazil, there were more animal companions than children in households, and that its human inhabitants took care of, and had conflicts with, different types of animals and plants (Baquero 2021c, 2021d, 2021a, 2021b, 2021e, 2021f). Building on this and other studies about multispecies health, in this chapter we will explore how the concentration of violence in urban peripheries affects the lives of multispecies collectives.

The problem of syndemics encompasses complex relations and processes—historical and ongoing—that include various modalities of violence, urban peripheries, and marginalized multispecies collectives. They are an ecological problem in which some human and other-than-human lives are systematically demeaned and disproportionately affected for the benefit of others.

When the promotion of multispecies health (Baquero 2021g; Baquero, Benavidez Fernández, and Acero Aguilar 2021) is concerned with marginalizing apparatuses that create symbolic and geographic peripheries, the ecological problem of violence is put into perspective, showing that epidemic processes are symptomatic of the oppression of multispecies collectives. We will see later that the promotion of multispecies health is about overcoming the pathological effects of marginalizing apparatuses that structure and operate the ecologies of violence.

The São Remo favela in the West Zone of the city of São Paulo is a periphery with which we have engaged to promote the health of multispecies collectives. In January 2019 the community living in households there had 8,457 humans, 983 birds, 745 dogs, 685 cats, and an unknown number of other living beings (Silva, Peçanha, and Gonçalves 2021). In previous work, we built a collective discourse on the experiences of the COVID-19 pandemic in the São Remo favela (Amorim et al. 2021). Through interviews with, and photos taken by the residents themselves, this community-based study portrays the experiences of 15 São Remo residents of the part of the community then living in the most precarious conditions. As part of the project, researchers produced a single document, organizing text and transcribed audio messages accompanied by photos that were representative of such experiences. Throughout the study it was continually updated, discussed, and approved by all participants. The resulting document contains the collective discourse and was shared with community members and published online. For the sake of participants' safety, most of the content regarding violence was excluded from the published collective discourse. Thus these types of violence make it difficult to identify, name, and draw attention to them. This results in the perpetuation of the dynamics behind the violence and the generalized lack of knowledge about the realities experienced in urban peripheries.

To contribute to the understanding of how violence aggravates and is aggravated by health emergencies triggered by infectious diseases and applying a multispecies perspective, in this chapter we put forward the idea of "ecologies of violence." In so doing, we aim to shed light on the entanglement that frustrates prevention and protection efforts in health emergencies: an interweaving of marginalizing apparatuses, modalities of violence, and health emergencies materialized in multispecies collectives.

EXPERIENCING THE ECOLOGIES OF VIOLENCE IN URBAN PERIPHERIES IN TIMES OF HEALTH EMERGENCIES

"There's so much going on here. . . . The people spend their days, running, turning the other cheek, every day getting up, facing struggle, there's always some obstacle. The pandemic is just one more obstacle that we deal with day to day for the men and women of the community, it's just one more of the big daily problems that we're going to face . . . actually, it is just one situation among thousands. Is it serious? It is! But there's millions of other serious stuff that happen to those who live in a peripheral community." "It's horrible to feel like nothing, . . . invisible, to see this bunch of mothers who are already there with a child, with a dog, with everything that's left, you know? In those tiny spaces, without water, we have to get water by the gallon to wash dishes and cook food" (Amorim et al. 2021, n.p.).

The above quotes are part of the aforementioned collective discourse, which was produced at a time when experts and mainstream media were telling the public to stay at home and wash their hands, disregarding the many who do not have a home or access to clean water. It was also the period in which new Brazilians were added to lists of international billionaires; banks beat profit records, and agribusinesses' exports set new records. All while poverty and food insecurity soared.

The collective discourse served to document and share what the pandemic living experience was like in a favela. The published document included only the collective discourse; it does not discuss the ecologies of violence. Based on what we deem the underlying processes of the published and omitted content of that collective discourse, in this section we draw some considerations that will better frame our approach to the ecologies of violence intensified by health emergencies in urban peripheries.

Urban Capitalism

The violence endured in urban peripheries is part of the very makeup of cities. In them violence is used to establish urban models that serve market interests, especially those in the real estate sector. To satisfy such interests, land is gained through the eviction and displacement of local peoples. Consequently, these populations, who are already marginalized by other social markers, are pushed out of the areas where jobs, cultural offerings, and quality housing are concentrated (Rolnik and Klink 2011; Rolnik 2001). Although at times there is a partial overlap between market interests and the quality of life experienced in some areas of the cities, when social policies are not in place, housing becomes a financial asset to be used for profit, not for the sustainability of cities nor the provision of good living (Rolnik 2013). As a product of this logic, urban peripheries are set on unstable land, prone to flooding, and where there are no housing ownership safeguards (Rolnik 2001). It is in these peripheries that, during infectious disease outbreaks such as COVID-19, we find the largest clusters of infection.

Housing conditions exacerbate the impact infectious diseases have on the health of those living in the peripheries. Population density, insufficient basic sanitation, and unhealthy indoor environments are compounded by labor exploitation, food insecurity, and on many occasions, long journeys in the overcrowded transportation that connect residential and workplace areas. As a result, stress, increased exposure to infected people, and comorbidities are a perfect storm for the contraction and transmission of infections, as well as for their severity and lethality.

The peripheral situation is a risk factor that is manufactured necropolitically, and not only by the violence of the aforementioned real estate market model. The privatization of health and the underfunding of the National Health System (Scheffer and Scheffer 2015)—another vector of structural violence—have intensified the vulnerability of peripheral subjects who are unable to pay for a health care plan with adequate coverage to meet the individual consequences of health emergencies. This situation is worsened by the lack of funding for education and assistance policies for children and adolescents, for the promotion of racial equality, or for efforts to combat violence against women, to mention a few (Instituto de Estudos Socioeconômicos 2021).

In this scenario, when stores cannot open and restrictions to prevent the agglomeration and mobility of persons are introduced to reduce the transmission of pathogens, unemployment grows and accentuates informal employment. Faced with having to make a choice between paying rent or buying food, the imperative of feeding themselves leads families to live in even more precarious homes, if not on the street. When people are forced out onto the street, new communities emerge in vacant lots and abandoned buildings. This is how the favela of Buracanã developed during the COVID-19 pandemic inside São Remo (Amorim et al. 2021).

Forced Displacement and Land Occupation

Amid the health emergency due to COVID-19, hundreds of shanties were built in Buracanã, occupying areas of steep terrain covered with rubble. These shanties were built on top of the rubble with all sorts of materials, including tarpaulins commonly used for walls and roofs. Many, however, had no roof for months (fig. 4.1). Consequently, people hung plastic sheets to protect themselves and their animal companions from the rain. Although shanties were built mainly over the first two months once the land occupation started, the replacement of tarpaulins with ceiling tiles took longer (see the colored plastic covers progressively replaced by tiles in fig. 4.2).

Materials, tools, time, and food were harder to find and purchase during a period of increased unemployment, mobility restrictions, and high inflation. Such circumstances explain the slowness of construction and why people had to spend nights with no power or water, while exposed to the weather, for several months. Without refrigeration or any proper storage structure, there is no possibility of storing perishable food or of preventing rodents and other synanthropic animals from contaminating the food. With the lack of a sewage system, the alternatives available increase the risk of infection. Using the bathroom in a nearby parking lot or a neighbor's bathroom in other parts of the favela increase physical contact in poorly ventilated community spaces. The construction of cesspools inside the shanties—small, with dirt floors and poor ventilation—is another alternative that, although reducing the usage of community toilets, brings about several other infection risks. Moreover, the risk and severity of fires is higher in locations where candles have to be used for lighting and gas cylinders are kept in inappropriate locations. Floods occur and insects proliferate in water puddles, and since the area was previously a landfill, it is swarming with scorpions that have to be collected daily (many are taken to an institute that produces antidotes,

Oswaldo Santos Baquero, Sara Cristina Aparecida da Silva, and Júlia Amorim Faria

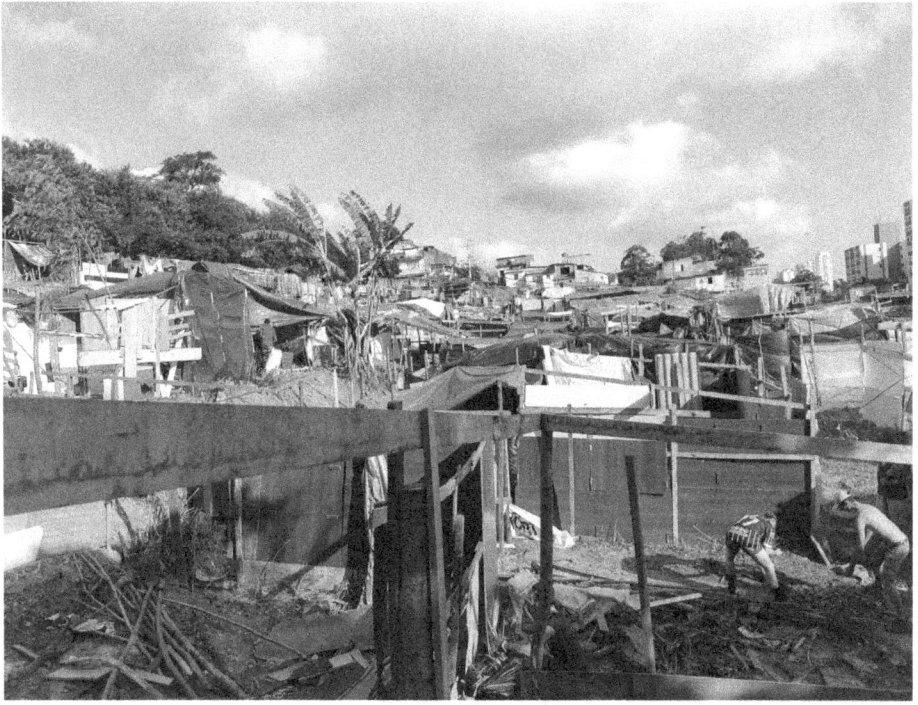

FIG. 4.1. Buracanã, occupation within São Remo favela during the COVID-19 pandemic. Photograph by Oswaldo Santos Baquero.

the rest are killed). Thus one health emergency precipitates others due to the ecologies of violence that undermine peoples' livelihoods.

The irregular occupation of land transfers the violence that triggers it to the collectives already occupying the same land. The irregular occupation occurs in empty buildings, vacant lots and environmentally protected areas (Rolnik 2001; Maricato 2019; Alvim, Kato, and Rosin 2015; Mesquita, Silvestre, and Steinke 2017). In this last case especially, local ecosystems are disturbed when the occupying multispecies collectives—made up of humans, pets, and the synanthropic animals attracted by human settlements—force the migration of different species already there. This process involves deforestation and increases morbidity, mortality, and therefore the suffering of plants and animals. The occupation of Buracanã, despite not being in an environmentally protected area and having a more central location, shows that these multispecies conflicts take place in urbanized areas, activating both violence and empathy toward other living beings: "We have to stay strong out of love for those people in the occupation, out of love for the animals that are already

FIG. 4.2. Progress in the occupation of Buracanã during the COVID-19 pandemic. Note the colored tarpaulin covers in the initial state of the shanties, gradually replaced by tiles. Yellow polygons delimit the occupied area, and dates have been added by the authors. Source: Mosaic elaborated by authors using Google Earth images.

there. We need a trap because there are many . . . animals. Like skunks that end up being dangerous and then die . . . trying to defend a litter or because we invade their space. We totally understand them and we don't want to kill them. We want to capture them and release them where there are bushes" (Amorim et al. 2021, n.p.; our translation).

Forced migration, furthermore, generates new problems in other times and places, including damage to bodies of water due to nearby occupation. In a scenario where basic sanitation services are privatized and their provision guided by profit rather than meeting basic needs, inequities in access to these services worsen and lead to the destruction and pollution of springs, hydrographic basins, and reservoirs that supply cities—and thus intensify the impact of health emergencies (de Sousa 2020).

In general, the occupation of environmentally protected areas is neglected, especially in comparison to the attention given to the occupation of idle buildings and land in more central areas of the city, which are more desirable to the real estate market. During the COVID-19 pandemic, the epidemic of unhoused people and the ensuing occupation of private properties also triggered an epidemic response in the form of threats of repossession (Nações Unidas 2020; United Nations 2020) and police violence (Chade 2022), especially in places coveted by the real estate market.

Unhoused Multispecies Collectives

The multiple forms of violence that lead to expulsions and occupation drastically reduce the subsistence options available. Those who live on the streets have to face health emergencies in conditions of extreme adversity and even the "help" that is available to them is itself imbued in vectors of violence and offers poor sanitary conditions. This is the case in shelters for people living on the streets, which are insalubrious, do not allow their animal companions, and are insufficient in number to meet the demand. During the COVID-19 pandemic, the Human Rights and Citizenship Commission of the municipality of São Paulo reported the existence of shelters with out-of-service bathrooms, clogged toilets, missing showers, and sinks with no faucets; mattresses with bedbugs and pigeon droppings; and overcrowded sheds where people lived alongside pigeons (SP2 2022). This is a clear example of structural violence, especially considering that the city government failed to use 30 percent of the budget planned for assistance and social programs for which it was meant (SP2 2022). Moreover, access to these shelters is restricted, especially when it excludes other-than-human animals. The affective bonds between humans and their animal companions are particularly important in unhoused situations, since they provide mutual support and care for humans and other animals alike. Despite this knowledge, and with only rare exceptions (Okumura 2022), Brazilian shelters do not allow these animals. Therefore, when faced with the impossibility of keeping their animal companions, multispecies solidarity makes many people opt to remain on the street (Singer, Hart, and Zasloff 1995).

Incarceration

The prison-industrial complex is a source and form of concentration of profit that exploits the supply of prison goods and services, including those produced by incarcerated human labor (Davis 2003). Since for some it is profitable to maintain and increase the incarcerated population, evidence for the ineffectiveness of prisons as a mechanism for resocialization and crime reduction is systemically ignored, much as the creation of alternatives guided by the fight against systematic violence that marginalizes and provokes criminal responses (Davis 2003).

For these reasons crime that takes place during health emergencies, which is caused by marginalization and committed to ensure access to resources, is conveniently used to incarcerate people and feed the prison-industrial complex in the name of public safety. Certainly the surge in crime is not motivated solely by extreme hunger and precariousness—especially if we consider the crimes committed by large capital holders targeting more wealth. Nevertheless, crime is not invariably the only way people face scarcity and, when associated with vulnerable contexts, is not necessarily committed by people driven by precariousness. However, the despair that health emergencies bring to the most vulnerable can lead them to commit criminal acts, a problem that should be answered by inclusion policies and the provision of decent housing conditions. Instead, this type of crime has been construed only as a threat to public security, leading to more police violence and persecution of peripheral subjects.

Although, overall, robbery and crimes against property decreased during the pandemic, food theft and murders by police increased (Folha de São Paulo 2020; Adaylton et al. 2022). During the first year of the pandemic, the number of murders resulting from police interventions in Brazil was the highest ever recorded, and as usual young black men were the largest group of victims (Bueno, Marques, and Pacheco 2021). Thus the greatest suffering for many mothers during the COVID-19 pandemic has been caused not by the viral disease itself, but rather by the disappearance, murder, or incarceration of their children—a suffering also shared by the children who lost mothers and fathers, sisters and brothers, or other dear relatives.

Thus prison configures itself as one more destination for these peripheral subjects during health emergencies. Carceral ecologies reflect the desire to build spaces devoid of life, except that of thoroughly surveilled and controlled humans. However, these are not purely human ecologies; they include microbiomes and sometimes synanthropic animals and "weeds." While detaching prisoners from other-than-human animals and plants is part of the punishment, gardens and animal companions provide limited possibilities for care and socialization in carceral ecologies.

Pathogenic agents are among those forms of life that move through the porous prison barriers, where overcrowding, poor ventilation, and limited access to water are the largest causes of mortality from infectious diseases (Sánchez et al. 2020). Control measures, like prohibiting visits and interrupting group activities (sports, school, educational and religious programs), ignore the structural causes of incarceration and serve only to prevent transmission to the outside of prisons at the expense of the detainees' psychological health (Sánchez et al. 2020). Although to an extent these control measures are better than nothing regarding disease transmission, without improving the sanitary conditions of prisons and, above all, without addressing the structural violence that drives crime and incarceration, they are far from an effective strategy to interrupt the dynamics of epidemic transmission.

Families and Household Environments

The structural violence affecting different areas of life is also exercised in another space—one typically associated with violence but not in its structural amplitude. The domestic space, in its household dimension, has been highlighted as a context of violence, usually understood and mediatized in the form of aggressions by one family member against another. These aggressions, both physical and psychological, presumably have also become epidemic during the COVID-19 lockdown due to greater home confinement (Dulius, Sudbrack, and Silveira 2021).

Home privacy and threats from perpetrators make cases of domestic violence difficult to detect. The fear of consequences for both those who consider seeking help and for other victims, or potential victims, at home discourage people from seeking shelter, and even more so from reporting aggression. Moreover, and once again, shelters' refusal to allow entrance to animals acts as a further barrier (Faver and Strand 2003; Cleary et al. 2021; Upadhya 2013). Since the fear of what might happen to their companions may dissuade victims from seeking help, the bond between humans and their animal companions works as a form of coercion in such cases.

The collective discourse on the experiences of the pandemic in the São Remo favela offers evidence for how this bond was a source of emotional support: "[The] contact with animals has become essential in fighting the virus, mainly because of the advantage . . . that they are companion animals, . . . that are by our side in moments of happiness, sadness, joy, and in good times. That's why we have to welcome them, give them more attention, since what they do for us is reciprocal" (Amorim et al. 2021, n.p.; our translation).

The violence triggered by the stress caused by marginalization and potentiated by the isolation and unemployment brought about by the pandemic is also referred to in the collective discourse: "With isolation, people end up getting more stressed, they harm animals and humans. They get stressed over not being able to leave the house, [. . .] being fired from work, they take it out on the animals, the people at home, this is happening a lot, you know?" (Amorim et al. 2021, n.p.; our translation).

Note the expression "more stressed," suggestive of a "basal stress," or in other words, the effects of the structural violence operated by marginalizing apparatuses. Thus understandings of domestic violence as being caused solely by familial problems and psychological dysfunction ignore the role of social reproduction in the structuring of families and the violence that occurs within and across them.

ECOLOGIES OF VIOLENCE

Now we turn to the ecologies of violence, derived from the ontological, epistemic, and praxical polysemy of a particular conception of multispecies health (Baquero 2021g):

> ontological, as it refers to material attributes, states, and processes experienced by marginalized multispecies collectives;
>
> epistemic, in the sense of an ecology of knowledge that problematizes and proposes health discourses and practices in favor of good living (buen vivir) for multispecies peripheries;
>
> and praxical because actions are informed by the knowledge that marginalization has pathological effects, and knowledge is produced through actions of health promotion in the midst of and against marginalization.

"Marginalizing apparatuses" is a key concept in the promotion of multispecies health. Through these apparatuses, margins are drawn to distinguish lives with intrinsic value and lives with instrumental value whose interests matter little or nothing. Those who draw the margins arrogate to themselves the prerogative of deciding who is worth more and who is worth less. In this way they legitimize biopolitical regimes to maintain their privileges by exploiting marginalized lives (Baquero 2021g). Marginalizing apparatuses do not act in isolation; they interact, structure, and instantiate the ecologies of violence. Apparatuses such as (neo)colonialism, racism, sexism, classism, speciesism, ableism, and ageism all interact and place some bodies, multispecies collectives, and territories on the "other" side of the margins.

The ecologies of violence are constituted by the interdependence among discourses, institutions, bodies, artifacts, and material-semiotic distributions and flows, in which marginalizing apparatuses establish symbolic and geographical peripheries, victimizing multispecies collectives in the process (Baquero 2021g). In other words, peripheries are ecological spaces on the outer side of biopolitical margins, inhabited and composed by victims of violence.

Although it is beyond the scope of this chapter to discuss why we are referring to multispecies health (which we previously called One Health of Peripheries) instead of One Health (OH), it is worth noting that the former is an explicit decolonial endeavor to promote multispecies justice and well-being. Intersectoral collaboration, transdisciplinarity, community participation, and optimization of human/animal/environmental health parameters are at the core of mainstream OH, which, after all, is concerned with a particular conception of multispecies health. However, it is a concept that reinforces anthropocentrism and pathological processes, framing humans in relation to but apart from animals and environments and subordinating other-than-human health to the reproduction of unjust and unsustainable modern lifestyles (Baquero 2021g; Baquero, Benavidez Fernández, and Acero Aguilar 2021). Moreover, mainstream OH, with its insistence on the added value of selected health parameters, can reinforce and legitimize unjust and overall unhealthy regimes. Slavery is illustrative: while the optimization of some health parameters of slaves was beneficial to them and brought added value for their masters, it did not abolish slavery; on the contrary, it reinforced it. Similarly, the OH signature is used to improve the performance of health indicators of otherwise marginalized multispecies collectives, for example, in large-scale industries that produce animal-derived commodities. In mathematical jargon, any local optimization is not necessarily directed toward global optimization, and therefore it is not necessarily beneficial for overall multispecies health. Of course, there are trade-offs, so that preventing or mitigating some forms of marginalization can simultaneously reinforce marginalization in other ways. However, it does not mean that the deconstruction of marginalizing apparatuses is irrelevant. On the contrary, such deconstruction supports the promotion of multispecies health when it is informed by decolonial, biopolitical, and Latin American health perspectives (Baquero 2021g; Baquero, Benavidez Fernández, and Acero Aguilar 2021).

Urban peripheries, as other peripheries, manifest historical processes in the ecologies of violence that systematically seek to erase the peripheral expressions that denounce it. Health emergencies intensify these manifestations and erasure attempts, intertwining preemerging and emerging circumstances in the generation and administration of suffering. Therefore what counts as violence, what and who is violent, and what and who can be a victim is a classificatory decision with profound ethico-political repercussions. This echoes a guiding set of questions to promote multispecies health: What is health, who can be healthy, who must be healthy, who can answer these questions, and who is heard by whom? (Baquero 2021e).

In the field of health, epistemic authorities exercise power through the knowledge they legitimize. The World Health Organization (WHO) defines violence as "the intentional use of physical force or power, threatened or actual, against oneself, another person, or against a group or community, that either results in or has a high likelihood of resulting in injury, death, psychological harm, maldevelopment, or deprivation" (Krug et al. 2002, 1084). This definition refers to the "power" to contemplate acts of omission and negligence. It goes beyond the immediate causes of injury, disability, or death to consider the effects of violence that are a substantial burden for the victims, even if considerable time has passed since the act occurred.

Indeed, the WHO's definition is comprehensive and may seem exhaustive at first glance. However, other perspectives suggest it needs revision. Lee (2019) believes that a definition of violence must make explicit the structural processes of oppression, as well as its potential to extinguish humanity and other forms of life, for example, through weapons of mass destruction. In considering this potential, a new definition would resize the importance of violence, placing it as a threat to humans and other species.

Structural violence is pervasive but also elusive because it uses historical erasure to legitimize systematic oppressions that guarantee prerogatives to the perpetrating social groups (Farmer 2004). Something similar can be said of symbolic violence, understood as the "capacity to impose the means to comprehending and adapting to the social world by representing the economic and political world in disguised taken-for-granted forms" (Richardson 2020, 57; Swartz 1997). Like structural violence, epistemic violence also operates by silencing the production of knowledge in the peripheries (Spivak 1998).

Another criticism of the WHO's definition concerns other-than-human victims. Although the status of "person" is not exclusive to human beings (Chan and Harris 2011), the aforementioned definition implicitly refers only to humans. The WHO, together with the Food and Agriculture Organization of the United Nations, the World Organization for Animal Health, and the United Nations Environment Program, is part of a global alliance that sees One Health as an integrated and indispensable approach to the planet's major health challenges located at the human-animal-environment interface (WHO 2022). Thus the WHO recognizes other-than-human animals as beings in which health is present or compromised and understands violence as an important health problem. However, it excludes other-than-human animals from its definition of violence.

Environmental violence also evades the WHO's definition. Environmental damage, and the suffering it causes, is a form of violence that is leading to the extinction of species and threatens the very existence of humans on Earth (Lee 2016).

In light of the ideological and systematic dimensions of structural and symbolic violence, and of the harm inflicted against other living beings and the environment, violence can be thought of as human action with the intention of causing suffering to other living beings or to oneself, or of benefiting from the intimidation, control, or exploitation of other living beings, through physical, verbal, behavioral, or ideological means that damage, directly or indirectly, occasionally or systematically, socio-environmental relationships or organisms, thus causing injury, suffering, death, or the extinction of species and ways of living. However, in addition to trying to improve this definition, one must consider, as Rodrigues (2021, 16; our translation) points out, that "[in] part, the problem of violence stems from identifying who forces you to name what is inside or outside the field of legitimate violence." Butler, in their discussion of the boundaries between violence and nonviolence, points out that "there is no quick way to arrive at a stable semantic distinction between the two when that distinction is so often exploited for the purposes of concealing and extending violent aims and practices. In other words, we cannot race to the phenomenon itself without passing through the conceptual schemes that dispose the use of the term in various directions, and without an analysis of how those dispositions work" (Butler 2020, 6). The conceptualization of structural, symbolic, and epistemic violence used to explain health emergencies, whether tuberculosis, AIDS, Ebola, or COVID-19, problematizes the way in which these violences have led to the emergence of epidemics and allow the resulting health emergencies to be used to ensure the privileges of the few to the detriment of the many (Sirleaf 2018; Richardson 2020; Farmer 2004; Wallace et al. 2020; Petteway 2022).

With this in mind, we can look at the question asked by Wallace et al. (2015, 70–71): "How, for one, does the World Bank or the World Health Organization approach outbreaks that originate with the very institutions on which the organizations depend for funding and legitimization?" These authors challenge the discourse of epidemic outbreaks by locating the hot spots in financial centers—New York, London, Hong Kong—rather than in remote areas and "wet markets," where pathogens spill over from other-than-human animals to humans. The agribusinesses and other extractive enterprises that are responsible for deforestation and the disruption of ecosystems that cause pathogens to spill over are, in fact, transnational oligopolies that accumulate profits in the Global North (Wallace et al. 2015).

The systematic omission of these processes of disease production in the mainstream epidemiological (and One Health) discourses has been argued to be a form of structural and symbolic violence (Richardson 2020; Farmer 2004). Such discourses omit these processes while highlighting a "natural history" in which infected individuals transmit pathogens to susceptible individuals (the more so with greater contact between them), with infection severity depending on the interaction between the pathogen's attributes and the host's immune system. From this perspective it is not surprising that the solutions proposed are genomic surveillance, early detection, and the introduction of timely transmission mitigation measures. While this is certainly desirable, it is impractical in many peripheral areas where the same colonial/capitalist processes facilitate the emergence of infectious diseases, a reality often ignored by epidemiological explanations. In other words, during health emergencies hegemonic definitions of violence continue to ignore the violence perpetrated by hegemonic groups. Thus the modalities of violence are codetermined, forming networks or ecologies of violence.

The interdependence of violence hurts the interdependence of the living beings affected by it every time they are configured and transfigured through the relationship with other creatures. Here, once again, Butler enlightens us by pointing out that the violated interdependence goes "beyond the dyadic human encounter, which is why [violence and] nonviolence pertains not only to human relations, but to all living and inter-constitutive relations" (Butler 2020, 12). Therefore violence is "an attack on persons, yes; but perhaps most fundamentally, it is an attack on '*bonds*'"; "it is not just other human lives, but other sensate creatures, environments, and infrastructures: we depend upon them, and they depend on us, in turn, to sustain a livable world" (Butler 2020, 16).

The shift from putting the emphasis on violent and abused individuals to the broader more-than-human entanglements in which violence is structured and materialized presupposes an ontological, ethical, and political change (Stanescu 2012). It is certainly not a change to detract from the act perpetrated, say punching a person. Yet it is one that seeks to understand these acts as an expression of historical and structural processes, making clear that they not only undermine human life (Butler 2020). As Stanescu says, "What is powerful is not what makes us unique, but what makes us in-common. What is exhilarating is not what individuates us, but rather what brings us together. This means that precariousness forces us to examine our connections, our methods and meanings for coexistence" (Stanescu 2012, pp. 576–577). Violence is a serious and complex phenomenon in which humans are not the only victims; they are part of precarious (in Butler's sense) and vulnerable multispecies collectives.

The ecologies of violence are an assemblage of discourses, institutions, bodies, artifacts, and material-semiotic distributions and flows in which marginalizing apparatuses establish symbolic and geographic peripheries, victimizing multispecies collectives in the process. Modalities of violence are interdependent, structured, and structuring and have pathological effects in multispecies collectives. An inquiry into them is an ecological exploration. In this chapter we argued that violence is a disputed concept, whose definition can legitimize the harm inflicted on certain living beings.

Health emergencies are intensified by and deepen the ecologies of violence, affecting some multispecies collectives more than others. In agribusiness, where other-than-human (industrialized animals) and human workers are victims of violence, pathogens and opportunities for zoonotic spillover appear. These zoonotic emergencies also occur in ecosystems violently invaded by agribusiness, extractivism, and urbanization. This is partly because the financial centers that structure networks of violence also structure networks of epidemic spread. One of the recipients and amplifiers of violence and diseases in these networks is the urban peripheries.

The COVID-19, globally spread by capital networks, reached and exacerbated the marginalization of Brazilian favelas. To highlight the materiality of the ecologies of violence in urban peripheries, we drew on our previous work, which included the lived experiences of residents in a favela in São Paulo.

By doing this we aim to bring greater visibility to the entanglement that frustrates attempts to prevent and protect against health emergencies: the interweaving of marginalizing apparatuses, modalities of violence, and health emergencies materialized in multispecies collectives. One of the ways to promote multispecies health is precisely through the deconstruction of the marginalizing apparatuses. It is important to recognize, first, that the ecologies of violence are inherent to these apparatuses' functioning and, second, to acknowledge the legitimation and concealment efforts that are being made to preserve iniquitous margins. In doing so the role of injustice as the basis of violence is highlighted, and the ecological configuration of these apparatuses is better understood. Thus actions to promote multispecies health are a way of confronting both the ecologies of violence and health emergencies, as they oppose the marginalization that reproduces peripheral epidemic configurations.

PART II

KNOWING ECOLOGIES?

CHAPTER 5

MODELS, EPIDEMIC THEORY, AND THE DECLINE OF THE ENVIRONMENT SINCE THE 1900S

Lukas Engelmann

The decades between 1890 and 1930 are often seen as the foundational years of "formal epidemiology." It is during these years, and particularly the period ahead of the World War I, where we now find a series of foundational myths, romanticized narratives of origin, and scenes of inception for what is known today as infectious disease modeling (Dietz 2009; Fine 1975; Heesterbeck 2002; Morabia 2013; Kucharski 2020). Over the twentieth century infectious disease modeling has gathered an astonishing authority in public health and governance.[1] With COVID-19, modeling, rather than more traditional forms of epidemiological thinking, could claim the authoritative status of trusted scientific expertise in the UK and elsewhere. The language of modeling has since become the vernacular of our epidemic times, and the making of models has advanced into a key technology for the securitization of health (Lancaster, Rhodes, and Rosengarten 2020; Rhodes and Lancaster 2020; Montgomery and Engelmann 2020; Hinchliffe 2020; Engelmann et al. 2022).

With the rise of mathematical approaches to epidemic curves in the early twentieth century, several currents in epidemiological theorizing shifted to the background. On the one hand, there were the increasingly futile efforts of Europeans to secure the identity and threat of epidemics to foreign spaces and populations, a decline of the steadfast belief in "hygienic modernity" that accompanied the decline of the great empires (Rogaski 2004). On the other hand, attributing the mass occurrence of disease to telluric, climatic, or even unsanitary conditions was challenged by the search for specific microbial agency, catalyzed by the industrious efforts of bacteriology. At the dawn of the century the environmental conditions that had dominated so much of epidemiological thinking across the nineteenth century might have had a continued presence in the observation and reports of epidemics, but the natural history of environmental determinism ceased to offer a convincing explanatory framework (Mendelsohn 1998; Anderson 2000; Amsterdamska 2004).

This chapter is concerned with the fate of the environment in epidemiological modeling at the beginning of the twentieth century. If topographical space and the environment have been a central aspect of understanding epidemics for more than a century, how were these components reframed in the reinvention of epidemiology as a formal science? If indeed ecological thinking has structured the perception of epidemic threats and risks since the early twentieth century, what are we to make of the flattened ecology in which the environment becomes a homogenized, uniform, and ahistorical variable? Here I ask how and why early modelers began to think of the environment in analogy to chemical laws of distribution. Has the flattening of space been a necessary condition to foreground the dynamics between pathogens and host populations, or was this the beginning of a more complex, and much more versatile, theorization of the environment as an ecological function? The environment of epidemic events, with its multiple material and social determinants, I argue, does not disappear into an imagined prototype of chemical mass action. No longer a determinant or causal component of epidemiological argument, the formalization of disease dynamics enabled the conceptualization of space as an interdependent variable, a function of both the pathogen as well as the population in the emerging epistemology of the "web of causation" (MacMahon and Pugh 1971).

This history is predominantly shaped in the circles of the Epidemiological Society of London, and infectious disease modeling remains at large an Anglo-American science. While the new epidemiologists at the heart of this story may well be seen as the originators of "armchair epidemiology," dedicated as they were to the production of theory and formula, they too were almost without exception enthusiastic agents of British imperialism. Their desire to craft a placeless and generalizable theory of epidemic dynamic should not be mistaken for a rejection of the racist foundations of epidemiology in the service of British overseas governance. On the contrary, a sound theoretical foundation of the science of epidemics was supposed, as Jacob Steere-Williams has shown, to support and strengthen what became at the time indeed an "imperial epidemiology" (Steere-Williams 2020). The early modelers' rejection of environmental determinism should thus not be read as a general disinterest in the material impact on host-pathogen relations, but was instead driven by an enthusiastic belief that all diseases exist independent of environmental determinants and that the key to their control lies in the dynamics between host and pathogen; the environment, in turn, became merely a secondary variable.[2]

The chapter begins with a reconstruction of the towering significance of space, place, and geography in the consideration of epidemics in the nineteenth century. The second section then follows the disentanglement of health and space in infectious disease modeling at the turn of the century and discusses the sudden emergence of the mass action principle as a metaphor and as a heuristic in epidemic theory. With the third section I consider the conditions under which the environment, and indeed Sydenham's "epidemic constitution," partially returns to epidemiology in the shadow of the 1918 influenza pandemic, and articulate some preliminary conclusions.

GEOGRAPHICAL PATHOLOGY

In the minds of Victorian geographers, sanitarians, and epidemiologists, the environment was a significant, perhaps a necessary component for the medical sciences since Hippocrates (Barrett 2000b; Rupke 2000). Indeed, in the nineteenth century "few issues stirred more public and scientific debates," Anne Buttimer (2000, 211) writes, "than those of health and environment." In these debates, the text of *On Airs, Waters, and Places*, attributed to the Hippocratics, took on a symbolic position as the ancestral narrative for most physicians, geographers, and historians throughout the nineteenth century. The premise of detailed local observations underlined the assumption that all diseases were a product of specific local conditions. For the ancients, observations included the forces and directions of the wind, the kinds of water, different temperature levels, as well as the direction of settlements. The Hippocratic text does indeed "float," as Valencius (2000, 8) argues, in most histories of health and environment "like a numerical constant" (8), offering a tale of origin and the authority and stability to foster a belief in the dependence of health on place.

Where the Hippocratics' consideration of the environment remained dedicated to the balancing of humors, post-Enlightenment thinkers in the nineteenth century aimed to redraw the contours of place within the project of a modern science. The second key reference to establish what would become medical geography, and later geographical pathology, is found in the work of Thomas Sydenham. While concerned with the Hippocratic heritage of environmental thinking, Sydenham also sought to establish concepts of disease specificity in the seventeenth century. Rather than assume a generalized miasmatic influence that would and could foster the development of illness, Sydenham advanced the localized idea of an epidemic constitution to infer "an orderly pattern to disease" (Barrett 2000b, 79). A particularly humid settlement, close to a river or surrounded by swamps, could be seen as a necessary condition for a secondary or tertiary fever. Not only were specific aspects, telluric influences, or geological formation tied to diseases, but the concept of an epidemic constitution also helped to advance a conception of diseases as distinctive and discrete entities. With the increasing interest in environs and their implication in matters of health also emerged the imagined assumption of being able to change and adapt the environment in the name of health improvement (Riley 1987).

Through most of the nineteenth century the conceptual relationship between environment and health was advanced under the rubric of geographical medicine. Esteemed milestones in this systematic approach to medical geography were Leonhard Finke's *Versuch einer allgemeinen medicinisch-praktischen Geographie* in the German context, Daniel Drake's *Systematic Treatise* on the diseases of North America in the 1850s, Arthur Bordier's *La géographie médicale* in France, and of course August Hirsch's *Handbook of Geographical and Historical Pathology*, published first in German but soon translated to rise to international recognition in the 1880s (Finke 1792; Drake 1990; Bordier 1884; Hirsch 1883; Valencius 2000, 9). French and American approaches to health and disease grew increasingly concerned with the control of territorial expansion. For France, Osborne describes a geographical "imperative" for considerations of hygiene as a way of correlating modes of life, human health, and places. However, Arthur Bordier also sought to expand the boundaries of a general field of medical geography, which he considered should include "anthropology, medical microbiology, and political utility," to then be placed at the forefront of the "colonisation scientifique" to advance the French imperial project (Osborne 2000, 44). At the American frontier, geographical thinking gathered a sustained following among settler-colonial immigrants and peaked with Drake's monumental mapping of the distribution of diseases between the Rockies and the Alleghenies (Drake 1990). Inclusive of meteorological, geo-

Lukas Engelmann

logical, and social determinants of health, Drake's work established calls for a therapeutic regionalism, fortifying a cosmology in which the management of human health was kept in close conceptual and geographical analogy to the place and land it occupied (Numbers 2000, 218). Only in the 1880s, and on the back of larger struggles about medical authority, as John Harley Warner writes, did the conviction that regional peculiarities inform health, assuming the "stigmata of inferior practice and antiquated thinking" (Warner 2014, 3).

The immensely influential German—or Humboldtian—school of geographical medicine was most likely prompted by the spread of cholera from India to Europe in the first decades of the nineteenth century (Broemer 2000). August Hirsch, whose seminal work received attention far beyond German sensibilities, emerged first as a fierce critic of the state of medical geography. He saw it as an ambitious field, plagued by a lack of data, poor scholarship, and little consideration for the immense challenges of establishing a sound science of geographical pathology and nosology. Most important, and in true Kantian fashion, Hirsch argued for the inseparability of historical and geographical approaches. Only an integrated analysis of the historical development and geographical distribution of diseases would eventually lead to the intended goal of establishing a scientific approach to the geography of diseases. Such a science would need to establish that symptoms of a disease were identical within a type of place to then secure a clear understanding of what the effect of the environment might be on that disease. With a focus on disease as the primary principle of the field, Hirsch envisioned his geography to rewrite a subfield of medicine beyond the confines of human anatomy: pathology, the science of the causes and effects of disease.

"The task of geographical pathology, as defined above, is to show how individual disease forms are generally distributed over the surface of the earth," writes Hirsch (as cited in Barrett 2000a, 108). The field's responsibility would be to identify "the geographically dependent factors (such as race, nationality, soil conditions, climate, social factors, etc.) that have to be considered essential for the occurrence distribution of individual diseases" (108). Geographical and historical pathology should thus assume a position among the medical sciences. Barrett warns it would be a mistake to associate Hirsch's reasoning with a miasmatic, or anti-contagionist conceptualization of disease (Barrett 2000b, 116). On the contrary, Hirsch deployed geographical reasoning to disprove explanations of disease distribution—such as miasma—that could no longer withstand the scrutiny of rigorous investigation.

Hirsch's work left a deep mark also in the circles of the Epidemiological Society of London before the advent of modeling. The science of geographical and historical pathology was greeted in Britain by a different but comparable endeavor to turn decades of dedication to the improvement of urban and domestic environments into a sanitary science. Worboys argues that the British sanitary science "was a synthetic subject that embraced statistics, law, engineering, chemistry, meteorology and geology, as well as medicine" (Worboys 2000, 111). Epidemiology was secondary to sanitary science, offering statistical accounts of increased occurrence of disease in space or social structures. As an obligation of the state and undertaken by the medical officers of health, sanitary science remained somewhat agnostic to causal explanations of various diseases and focused instead on the preventable nature of particular conditions, "regardless of their conceptualization as miasmatic, infectious, contagious, epidemic, occupational or nutritional" (112).

For the case of one of the most significant diseases in the history of epidemiology, Steere-Williams reminds us that the engagement with typhoid, the "filth disease," throws a bright light on the continued significance of the environment in the development of even modern British epidemiology (Steere-Williams 2020). The emerging traditions of epidemiological outbreak investigations that accompanied the filth disease in Britain entangled the place of disease not only as an abstract aspect of classification, but inscribed it as an experiential and essential element of epidemiological reasoning. From sanitary concerns about water pumps and hygienic accusations against milk to the recurring anxiety about pathogenic matter in the soil, the Victorian epidemiologist had to travel to the site of an outbreak and establish a sense of the material conditions that connected cases and yielded to patterns and clusters before inferences about causes could be made. Even in the face of the growing significance of bacteriological science, epidemiological methods on the ground maintained primacy in the consideration of typhoid fever as well as in cases of plague and even cholera (Engelmann 2018; Lynteris 2017; Whooley 2013). While the bacteriological revolution surely shifted the focus of conceptualizing the infectious disease toward the transmissible pathogen, the environment continued to inform and structure epidemiological knowledge as well as government actions.

The narrative of the rise of the laboratory is often simplified as one in which bacteriologists began to claim an exclusive authority over the domain of infectious diseases via the identification, cultivation, and in some cases, the successful vaccination against pathogens. A newly unified germ theory was suddenly available, so the story goes, to overturn long-held beliefs, to modernize approaches in the measurement and visualization of specific diseases, and to replace traditional sanitary practices with interventions led by science. Many historians have since invested in a more nuanced and considerably more complex narrative of the transition from a geographical to a laboratory framework. For Worboys (2000) this transition implied the loss of more inclusive approaches, as they were characteristic for the sanitary sciences, to a more exclusive engagement within the tension of pathogen and (individual) immunity. For Graham Mooney (2015, 6) the "remediation of the public environment was never abandoned, but there was a shift away from the preoccupations of the sanitary movement toward the medical management of biologically dangerous individuals." Olga Amsterdamska, when seeking to delineate the scientific profile of epidemiology around 1900, remarked that epidemiologists by and large remained less impressed by bacteriology than commonly assumed: "Most British epidemiologists reacted to the advent of the germ theory of disease neither with hostility and rejection nor with the belief that the new theory implied a subordination of their own findings to those of the bacteriological laboratory" (Amsterdamska 2004, 486). This included consideration of the environment, which neither suddenly disappeared from reasoned arguments about epidemics, nor did epidemiological thinking descend exclusively into the negotiation of the binary of human host immunity and pathogenic virulence. The question is then how space came to be reconfigured from a topographical environment into a set of topological vectors to inform epidemiological reasoning and to become more than just a remnant of Victorian dedication to infected soils and social destitution.

COMPLEX EPIDEMICS AND THE DECLINE OF THE ENVIRONMENT

For Linda Nash the rhetoric of the "new public health" in the early twentieth century, driven by enthusiasm for germ theory, was accompanied by a much narrower definition of health. As a consequence, the environment and the human body had been disentangled and had become the subject of discrete scientific projects. With germs envisioned to reside in the human body, negotiating the occurrence, distribution, and dynamic of infectious disease became almost inescapably submerged into the binary opposition of population and pathogen. Scientists, physicians, and epidemiologists alike hence disregarded the environment as a primary concern and shifted it to the periphery of their causal inferences. "Their models seemingly, and conveniently, resolved the tension between modernization and health by exonerating the landscape from any independent role in disease" (Nash 2006, 6).

This almost exclusive duality of pathogen and population informed much of the new advances in epidemiological theory at the turn of the century and enabled the emergence of mathematical modeling. Most of the early modelers and their theories circulated around population ratio and pathogenic virulence. They all, to varying extents, grappled with fitting epidemic curves to the statistical constants developed and popularized by Karl Pearson (Brownlee 1906, 485), or in some cases sought to develop novel theories to generalize infectious disease dynamics beyond historical circumstance. Comparing and contrasting historical records of measles, cholera, smallpox, plague, and other diseases to normal curves suggested a lawlike behavior, but the challenge remained to explain and to extrapolate the causes of any apparent regularity with regard to advances made in the sciences of bacteriology and immunology.

Ronald Ross, known widely for his work on malaria in British India was likely the first to develop a mathematical theory of infectious diseases without prioritizing a specific pathogen. His "a priori pathometry" sought to make assumptions about the spread of disease without reference to specific ongoing or past epidemics (Ross 1916; Fine 1975). His formulaic work was not only independent of the grand tales of past epidemics or the ongoing politics of quarantine at his time but was also indifferent to any material impact from locality, and operated instead—like most mathematical epidemiology—in a generic and uniform environment.

One reason for Ross's irreverence toward the topographical environment might not be found in what his decadelong research tried to prove—that mosquitoes are a vector for malaria in humans—but what he tried to disprove, namely that malaria was to be conceived as a disease of a specific kind of place. Ross expands in his Nobel prize lecture that malaria had been the paradigmatic "*endemic* disease—that is, that it does not easily spread from man to man independently of locality as do for instance, small-pox or plague; secondly that it adheres especially to warm localities where there is much stagnant water, such as marshes" (Ross 1905a, 29, emphasis in original). While developing his experimental work in various locations across British India to demonstrate the validity of the mosquito as a vector, Ross grew increasingly skeptical of topographical space as the widely assumed driving factor.

His closest competitor, the Italian malariologist Amico Bignami, might have agreed on the principle of transmission by mosquitoes but remained strongly attached to a saprophytic origin of the disease. Mosquitoes, so Bignami posited, acquired the parasite from the ground before infecting human populations: "They bring the poison from marshes to man" (cited in Ross 1905a, 46). Ross, on the contrary, began to develop the theory that the mosquito was merely an intermediate host, receiving the parasite from humans before infecting other humans. Ross became, as Warwick Anderson (2021, 171) recently argued, "obsessed with insect-human interactions—with measuring malaria transmission entomologically, by mosquito density—at the expense of any nuanced understanding of the sociality and cultures of human populations." Perhaps through extrapolation from this work on malaria, his mathematical work on a generic epidemic was almost exclusively shaped by mathematical concerns about population ratio and density.

If we turn to Ross's later publications, there is little explicit reference to the question of the environment at all. If anything, space is everywhere implied but rarely theorized or implicated in his mathematical reasoning. Ross (1905b) rephrased geographical space as merely a logical function, a topological relation of the interactions between insects and humans. In a speech subsequently published in *Science*, "The Logical Basis of the Sanitary Policy of Mosquito Reduction" (Ross 1905b), he gave perhaps his most explicit account of locality. To develop a principled basis for sanitary policy, any approach to the reduction of mosquitoes had to depart from a set of general observations, which should be formalized independent of specific local circumstances. While general mechanics of the population dynamics of mosquitoes were understood—for example, breeding cycles and average life span—the influence of local specifics could not be inferred from any experiments or measurements. While it was known that standing water offered a breeding ground, it remained unclear how that impact could be measured or calculated. "We cannot anywhere state the exact number of mosquitoes to the square mile or yard, and we cannot, therefore, accurately gauge any local decrease which may have resulted from operations against them" (Ross 1905b, 690). Given the absence of experimental means to determine environmental influence, Ross turned instead to "strict logical deduction" (690). To calculate the effect of actions designed to reduce the population of a living organism such as a mosquito, he turned to a thought experiment: "Suppose that we have to deal with a country of indefinite extent, every point of which is equally favorable to the propagation of gnats (or of any other animal); and suppose that every point of it is equally attractive to them as regards food supply; and that there is nothing, such for instance as steady winds or local enemies, which tends to drive them into certain parts of the country"(692). From here, with the assumption of a uniform distribution of mosquitoes within an abstract but uniform environment, Ross could begin to develop his characteristic mathematical approach to calculate how measures aimed at the destruction of mosquitoes compared against measures seeking to remove the conditions for their reproduction. With this thought experiment, Ross defined the spatial coordinates required to arrive at his mathematical reduction, while at the same time framing this spatial construct as an obvious piece of fiction. The topography of space, the arrangement of natural or human-made features of a landscape or environment were no longer part of the considerations in his epidemiological style of reasoning. In turn, his formal reduction allowed him to draw pathways for intervention and thus develop a vision for the control of malaria.

As a result, Ross theorization concluded that epidemic phenomena could and should be understood as the inverse relation between infected and uninfected individuals in a population. Assuming the primacy of population dynamics over external, material factors, he focused on assumptions about recovery rate or acquired immunity. But importantly, as Fine shows, for Ross's mathematical work to later exceed the realm of infectious diseases, he focused on other determinants of population structure as well, such as "births, deaths, immigrations, and emigrations of affected and unaffected individuals" (Fine 1975, 549),[3] laying the groundwork for a general theory of contagion independent of place, pathogen, or politics (Kucharski 2020).

John Brownlee, a statistician and student of Karl Pearson, was less concerned with the intricate dynamics of populations and instead advanced a mathematical focus on the infectious agent. In 1906 he mounted a strong argument that neither the number of susceptible hosts nor differences of place and time could account for the regularity observed in epidemic curves (Brownlee 1906). Indeed, he based his work directly on the curves William Farr had fitted to smallpox outbreaks since the 1840s. Given, however, that Farr's mathematical description did not illuminate the causes of observed regularity, he complained that it remained a "law without a reason" (Brownlee 1909, 245). Equipped with scientific approaches to germ theory, it was now time to offer a mechanistic explanation for what Farr had merely observed.

Brownlee sought to achieve nothing less but the redefinition of the principle of infectivity on mathematical grounds. He observed that a range of historical outbreaks could not only be fitted to curves but that these often resembled the shape of a *normal* curve (now commonly referred to as a Bell curve). Equipped with the statistical training derived from Pearson, where these curves were known as "type IV," this revelation "further convinced Brownlee that Farr's method contained some fundamental truth as to the mechanism underlying epidemic processes" (Fine 1979, 350). Supporting this theory, Brownlee argued implicitly and explicitly that location or the type of disease should have only negligible impact on the shape of the epidemic curve. His examples to demonstrate the generic and recurring character of epidemic curves included the 1665 great plague in London, measles in Glasgow in 1808, and smallpox in Gloucester in 1895–1896, among other outbreaks. Even

Lukas Engelmann

sanitary measures and a varying density of occupation (contrasting towns and villages) might only have led to a different "amount of disease in a given epidemic" (Brownlee 1906, 492), but always approximated the same normal curve with a steep incline and symmetrical decline. He concluded that any uniform intervention, as sanitary improvements might offer, can indeed diminish the severity of an epidemic and control the impact of the disease, but would not have "perceptible effect on the form of the curve" (492).

Brownlee's main argument was that such normal curves could not confirm a key assumption at the time; most bacteriologists assumed that the infectivity of pathogenic organisms was constant. If this were the case, Brownlee argued, curves would have looked different. If the changing amount of susceptible hosts in a population was the only and the determining factor—as Ross would argue—the epidemic curve would suddenly drop off, rather than exhibit a symmetric shape of growth and decline. The result would be a left-sided skewing of the curve, which he found to be inconsistent with the data he consulted. Instead Brownlee assumed that changes in virulence seemed to be the most probable culprit, and the shape of the curve demonstrated that the pathogen loses infectivity until "the end of the epidemic at a rate approaching to the terms of a geometrical progression" (Brownlee 1906, 516).

In the same lecture Brownlee also wondered if any such law, as described here for the dynamics of an epidemic over time, might be applied to its distribution in space. Rather off the cuff, he began to formalize a principle of spatial distribution. This law was elsewhere formalized as the "mass action principle," which the mathematician Fine boldly identifies as one of the "most important themes in all mathematical epidemiology" (Fine 1979, 348). With the growing influence of mathematical approaches to epidemics, the question of space became increasingly defined in terms of an "equilibrium system." Heesterbeck (2005) assumes (without archival evidence to support his inference) that the works of Ross, Brownlee, and William Hamer were all written with awareness of ongoing theoretical developments in physics and especially in chemistry. Two Norwegian scientists, Maximilian Guldberg, a mathematician, and Peter Waage, a chemist, had formulated the mass action principles as a law of chemical reaction kinetics. The law was received with much excitement around Europe in the late nineteenth century. Guldberg and Waage had observed that the rate of a chemical reaction is directly proportional to the activities of reactants. Or, in other words, that the relationship between a product and its reactants can be expressed in an equilibrium constant: $AB + CD = AC + BD$ and this constant state of equilibrium (k) can be predicted at any stage if the volume of reactants is known.

Brownlee (1906, 508) thought that a similar law could be easily defined for epidemiology: "Given a certain amount of infection in a limited space in the midst of a uniformly distributed population, it seems natural to assume that the chance of any individual coming into the zone of infection will approximate to that given by a normal probability surface of which the maximum corresponds to the area in infection." Incidence of a disease was understood to be expressed in the rate at which infected and susceptible hosts meet, which in turn Brownlee saw proportional to the product of the spatial densities of each of the respective populations. In other words, an epidemiological mass-action principle assumed that population could be conceptualized as a homogenous mass, distributed with continuous density in space to formalize a plausible epidemic curve. Heesterbeck, in his appraisal of the mass-action principle in the history of epidemiological science, implies that Ross, Brownlee, and others at the time assumed the population to be a "well-stirred" system (Heesterbeck 2005, 101).

What emerges here in rudimentary form and without explicit reference to natural laws developed in chemistry, came to structure the work of two further British mathematical epidemiologists at the time: William Hamer and Anderson McKendrick. Hamer, in his early publications on periodic fluctuations in the epidemic curves of measles, argued that neither variations in the pathogen nor the susceptibility of the host might be implicated, but rather that the regularity of such outbreaks was to be explained by reference to "a changing ratio between the numbers of susceptible hosts and the 'cases' able to infect others in a given population" (Amsterdamska 2004, 489). As Heesterbeck argues, Hamer thus imagined community as a "homogeneously mixed population with a constant force of infection," to develop a so-called proportionality constant "as the inverse of the endemic susceptible" (Heesterbeck 2005, 89). Interestingly, Hamer refused to believe that regular fluctuations of an epidemic curve would need to be attributed to either the properties of a pathogen or to a variable susceptibility in a population. His curve, as Hamer wrote (1906, 734), "sufficiently indicates that an epidemic may come to an end despite the existence of large numbers of susceptible persons in the population, merely on a 'mechanical theory of numbers and density,' and that the assumption of loss of virulence or infecting power on the part of the organism is quite unnecessary." As J. Andrew Mendelsohn (1998, 317) puts it, "Hamer's model could not be resolved into its parts" as the curve was not the result of a shift in the properties of bacteria or hosts, but the result of a "change in the *rate* of infection and the *density* of the interactants." While Hamer laid out some of the assumptions resembling the mass action principle, his hypothesis lacked

Lukas Engelmann

the required theorization of how infection occurs, how frequently individuals make contact, and how circumstances, such as spatial elements, impact the frequency of contact and infection (Heesterbeck 2005, 90).

Anderson McKendrick, a Scottish physician and acclaimed pioneer of epidemic theory, was perhaps most explicit in the application of the mass action principle to epidemic theory. He had served under Ross in various campaigns against malaria in India and Sierra Leone and shared his enthusiasm for the application of mathematical reasoning to epidemiological problems. He too was drawn to explanations for the regulation of epidemic curves to be found in the homogenous mixing of infected and uninfected individuals. McKendrick (1912, 59), however, focused on the question of "collision" in a paper from 1912, read before a conference on malaria. It was his intention to define k, the equilibrium constant, as "a factor which measures the chance of infection," including the degree of dispersion of individuals and the degree of intercourse. His theory was supposed to explicate the chance of transmission, through which an equilibrium constant between infected and susceptible individuals was reached (see also McKendrick and Pai 1912). Throughout the works of McKendrick and his colleagues, this utilization of differential equations to approximate the dynamics of infectious diseases quickly became a more or less standardized practice.

In the UK it was eventually Herbert Edward Soper, who rendered the mass action principle into an explicit foundation of his epidemic theory. Although he published his work only in 1929, it spoke directly to the line of reasoning initially developed by Ross, Hamer, and McKendrick. Like McKendrick, he assumed the chance of infection to be ultimately governed by the ratio of infected and susceptible individuals. But he went one step further and assumed "that a process analogous to 'mass action' governs the operations of transmission and that, other things equal, the number of cases infected by one case is proportional to the number of susceptible in the community at the instant" (Soper 1929, 37). As Heesterbeck remarks, Soper was quite satisfied with the synthetic epidemic that resulted from his mathematical work, but as one of the first modelers who sought to corroborate and adapt formula in confrontation with data from outbreaks,[4] he struggled to fit his theory to the data in all places. His curves matched well the data from a measles outbreak in London, but he could not make sense of the shape of the curves derived from measles outbreaks in Glasgow between 1888 and 1927 (Heesterbeck 2005, 99). Ultimately, Soper came to question the analogy between liquid dynamics in chemistry and the spatial factors governing the distribution of infectious disease: "In a liquid the intimate uniformity of the mixture and the conditions of intermingling and collision are likely to be more law-abiding than are similar traits in a community of persons" (Soper 1929, 54).

To mathematical epidemiologists, venturing to establish a new scientific field, the chemical equilibrium, as formalized in the mass action principle, was more than just a metaphor. In the work of Ross, Hamer, McKendrick, and Soper, it served as a vehicle to bracket the consideration of space and place from the development of lawlike representations of epidemics. However, neither the inception of bacteriology, nor the emergence of population in statistics had delivered sufficient explanatory models, but neither had they rendered the geographical space fully obsolete. Instead, the mass action principle enabled the reconfiguration of space to supply two conditions for a mathematical epidemiology to become reasonable. On the one hand, the elevation of epidemic theory into the realm of laws and principles comparable to the natural sciences required the epidemic to be lifted from the environment into an object, characterized by predictable dynamics that occur—to some extent—independent of place and time. Modelers wanted to move away from an epidemiology a posteriori, from history as much as geography, to devise a scientific concept dedicated either to a radical a priori pathometry (Ross) or at least to the development of theories and models, which could be continuously adapted, refined, and rebuilt to fit empirical data (Soper). On the other hand, the mass action principle enabled a rethinking of the meaning of place and space to conceptualize the environment within a rather narrow and flattening scale of more or less homogenous states of being. The locations within which epidemics emerged were no longer analyzed for their determination of epidemic circumstance but were compared against the new implicit norm of enabling or prohibiting a continuous equilibrium in the flow and density of individuals and their chance of contact. In the case of Soper, and indeed of most interwar epidemiologists, this thinking quickly turned into an obstacle. The topographical space and its inhabiting population had, after all, little in common with the properties of combined liquids. While Heesterbeck pointedly remarks that "systems adhering to mass action are, in a sense, well stirred" (Heesterbeck 2005, 101), societies and their social and material conditions remain too often structured by a myriad of historical, local, and political obstacles.

RECOVERING PLACE, REPHRASING THE EPIDEMIC CONSTITUTION

In his book *Epidemiology: Old and New* (1928), William Hamer revised the efforts he and his colleagues had mounted to devise a "theory of the epidemic wave" (Hamer 1928, 13). Instead, since 1919, he argued, a

Lukas Engelmann

"considerable 'Sydenham revival'" was in progress (13). Hamer, as much as his colleagues M. Greenwood (1919), Francis G. Crookshank (1922), and E. W. Goodall (1927), had begun to rethink the validity of Sydenham's consideration of the epidemic constitution. The obvious reason for this sudden return of place and space into epidemiological reasoning was the shared experience of the devastating influenza pandemic of 1918–1919. Both Amsterdamska and Mendelsohn attributed a significant shift in epidemiological thinking to influenza. Epidemics had suddenly become a lot more complex, while uncertainty about the drivers of epidemic dynamics had drastically increased. To its witnesses, the influenza epidemic had exceeded all historical precedence, and its devastation had been beyond predictions and assumptions. Was the exaltation of the unknown germ driving influenza a one-off event, a random chance event, as the epidemiologist Arthur Newsholme wondered, or did the germ's virulence increase due to some "lawful process of bacterial modification," as Greenwood argued (Amsterdamska 2004, 492)?

The sudden global eruption of influenza required explanations, which were difficult to achieve. After decades of theorizing the relationship of pathogens and populations as systems of chemical (im)balance, the reason for a sudden outbreak could hardly be attributed to external factors or to freak chance. Mendelsohn (1998, 306) argues that "the redefinition of epidemics, from invasion to disturbance of equilibrium, reflected a broad shift in disease patterns: namely, the replacement of Western experience of epidemics and pandemics as exotic plagues such as cholera, plague, and yellow fever, by the experience of epidemics and pandemics as arising from within—cerebrospinal meningitis, poliomyelitis, encephalitis lethargia, and epidemic influenza." Influential figures like Greenwood, whose biography linked long-standing skepticism about bacteriologists' explanatory models with eugenic sympathies, could now advance their interest in a recovery of Sydenham's epidemic constitution as an explanatory framework.

Greenwood turned to Sydenham to emphasize that while some epidemics might occur "perfectly uniform in their evolution" others are "variable from constitution to constitution" (Greenwood 1919, 59). Theorizing constitutions enabled the explanation of multiple epidemic phenomena emerging at once—a common view of what influenza was at the time—and it introduced epiphenomenal factors into the epidemiological algebra to explain long-term transformation of epidemics as well as local variation. While still vague in its implications and rhetorically distinguished from sanitary status or, for example, water quality, recovering the epidemic constitution shifted the focus of action toward the material and social world: "The moral is that a general consideration of the facts of human life, the slow changes of normal social evolution, the drastic changes enforced by recent events may be of as much importance as an intense scrutiny of the specific vera causa, the a, b, c of the bacteriologist" (65).

For a brief time epidemiological modelers had left the confines of the material world behind to instantiate the epidemic as a novel object of scientific reasoning, one whose regularity and lawlike behavior exceeded the specificity of pathogens as much as it escaped the determinism of constitutional disposition and environmental influences. With Warwick Anderson one could place this enthusiasm for an "anthropomorphic mobilization of pathology" (Anderson 2000, 147) in a period in which the consideration of environmental determinism was attached to the nostalgia of natural history. This nostalgia left little room for the modern project of reforming or circumventing the constraints of nature. With the decline of the environment in mathematical epidemiology, a vision of potential control and large-scale intervention seemed more tangible. Ross's mathematical work on malaria was meant to transform the global burden of the disease and thus to remake and change environment. The example of the malaria- and yellow-fever-plagued construction of the Panama Canal in 1904 speaks to the validity and the power of this vision, as William Gorgas attributed much of his success in the control of diseases to Ross's work (Coleman-Jones 2010). However, the example of influenza and the shock of the pandemic also demonstrates the gullibility of epidemiologists, when lawlike phenomena with little or no relation to history and geography assume autarchy. The return to Sydenham and the revival of a modern thinking of epidemic constitution put mathematics into place to reintroduce the ever-present question of the social as an incommensurability.

CHAPTER 6

"CLICK A BUTTON AND . . . SEE"

THE PERFORMATIVE POWER OF SYNDROMIC SURVEILLANCE IN UNDERSTANDING EPIDEMIC ECOLOGIES

Henning Füller

You shall know them by their fruits.

—Matthew 7:11

This chapter engages with current conceptions of epidemic ecology by considering their expression in current public health surveillance infrastructure. It draws on interviews with users of a recent innovation in public health monitoring to examine the technopolitics of such systems in shaping epidemic ecologies. This innovation is an indicator-based early-warning system for public health events, termed a *syndromic surveillance* system, introduced in the United States to monitor public health. Such additional health monitoring had been promoted shortly before the terrorist attacks of 9/11, and it received strong political support in their aftermath. Several technical applications of syndromic surveillance have been installed since 2001 in several US states, and an integrative system has also been established at the federal level. Interviewees in Washington, DC, Maryland, and Virginia used the Electronic Surveillance System for the Early Notification of Community-Based Epidemics (ESSENCE).

The interest in such a technical tool and the effort of financing and installing the infrastructure, training personnel, and establishing new daily practices at health departments arise from the securitization of health and a new paradigm of homeland security following the 9/11 attacks (Fearnley 2008). But the installation and use of such a system are not only an expression of changing paradigms and political preferences. Socio-technical infrastructure can also be understood as politically performative.

Recent debates in science and technology studies have called for greater awareness of the performative character of technology. The social relevance of technology is not limited simply to that of a passive object. The manifold interdependencies of user and tool are underlined in many current approaches of science and technology studies and anthropology. It is even proposed to understand this interdependence in a symmetrical fashion and to grant technology a generative capacity of its own in socio-technical relations (Pfaffenberger 1992; Sneath, Holbraad, and Pedersen 2009). "Technologies should be understood as both produced through culture and as productive of culture" (Lock and Nguyen 2010, 23).

I approach the example of syndromic surveillance: as a concrete socio-technical infrastructure that is politically performative in both grounding and shaping a current understanding of public health. The editors of this volume explicitly encourage such an approach to take tangible practices and institutional settings (see the introduction) as the analytical vantage point.

The need for the installation of additional health monitoring infrastructures such as syndromic surveillance has been amply demonstrated by a changing global environment and corresponding emergence of new health threats. New emerging diseases (Lederberg, Shope, and Cole 1992), most notably COVID-19, demand the strengthening of technical surveillance capabilities, early warning systems, and preparedness. What is important here, and this is also the premise of this volume, is to be aware of how these responses and their specific operationalizations alter our understanding of health and related concepts. An emerging disease worldview also acknowledges the interdependence of health and the environment. But how does this growing awareness of epidemic ecology translate into policies? Thinking about health and disease with an ecological framework today is often rendered through a pathogeno-centric, infection-control-oriented focus. The current way of thinking about health is strongly associated with an epidemic triad of environment–host–agent and a shift toward infection control, surveillance, and health security. This is especially apparent in current global health policies (Lakoff 2012, Weir 2012).

The problem of disease ecology is not a given, but it can be fruitfully considered as a problematization, a specific way of ordering and sorting the world according to a specific understanding of the environment–host–agent triad. This problematization is not only an issue of epistemes and meanings but is also practiced and established through tools, institutions, architectures, and databases. "Things are only objectivizations of determined practices," as Paul Veyne (1997, 159) summarizes in his relational reading of Michel Foucault. This perspective is applied here to answering the question: How is syndromic surveillance, as a certain set of determined practices, an active part of formulating the problem of disease ecology?

In the next section I describe in more detail how syndromic surveillance has been funded, implemented, and established as an innovation in public health monitoring across the United States. This innovation has been implemented in response to a shift in how emerging diseases and the global environment are reconsidered as health threats. Then I detail the technopolitics of syndromic surveillance. This involves engaging with the set of thoughts, modes, and practices that become objectified within this system of health monitoring. Finally, this specific rendering of epidemic ecology, as it is understood and established through sociotechnical arrangements, has certain effects and consequences. The last part discusses some of these political effects.

SYNDROMIC SURVEILLANCE AS AN ANSWER TO WHAT?

What is today called the National Syndromic Surveillance Program has evolved over the last two decades from several pilot systems tested and employed first at US military hospitals and later at several state health departments (Lewis et al. 2002). One feature common of all these predecessors and the current nationwide system, and the reason for the "syndromic" label, is a turn away from monitoring diagnosed cases. Instead, monitoring uses a diverse pool of nonspecific data, often only indirectly related to health issues as such. The core idea is to use a breadth of data accessible in near-real time for nonstop monitoring. Automated pattern recognition algorithms constantly scan through these data to detect unusual patterns and possible health threats (Henning 2004).

In the system applied in the United States, a central source of these near-real-time data are electronic health records of what are termed chief complaints, gathered at hospitals, but other nonspecific but health-related information is also collected through the system, including school absenteeism and pharmacy over-the-counter sales. The core feature is that these incoming data are automatically sorted and classified to enable nonstop algorithmic monitoring. The stream of data is sorted and compared to previous days, and any unusual spikes are then flagged automatically.

The main promise of this syndromic approach is to be faster in recognizing unknown threats. Such nonspecific public health monitoring has been conducted for a long time in several ways, but mostly based on archival data sources for population statistics such as birth rates, health insurance information, and health surveys. Syndromic surveillance instead uses nonspecific but timely data to detect possible clusters of infections or other health problems as they develop, without waiting for a medical diagnosis.

Generally, near-real-time surveillance of health issues is an established routine and an important part of public health monitoring that relies on a reporting requirement for practicing doctors. A General Practitioners' Sentinel System was established in the UK and the Netherlands as early as 1967 (Declich and Carter 1994), but such monitoring is routinely disease-specific and asks the participating sentinels to call in any evidence matching a predefined catalog of health concerns. The innovation of syndromic surveillance is to be able to flag any spike in the data without a built-in preference for a certain disease or condition. Issues are detected regardless of the source, be it an unknown threat such as a bioterrorism attack or a currently unknown disease (Morse 2012). In sum, the system promises greater speed than nonspecific, ex post facto monitoring of archival data and greater sensitivity than disease-specific, near-real-time monitoring.

Individual showcase examples of the benefits of such monitoring are rare. The H1N1 outbreak in the United States in 2009 is often cited as such. The early warning signs gathered through the daily data stream of syndromic surveillance allowed epidemiologists to model a rising incidence some days in advance of other sentinel-based monitoring. However, the validity of these numbers has been questioned (Stoto 2012). Respondents in a case study in Maryland consequently downplayed the biosurveillance application and instead emphasized the benefit of syndromic surveillance as an additional tool for a broader situational awareness of the state of public health (Füller 2022).

Syndromic surveillance in the United States began in the late 1990s. As the internet was established, the promise of rapid and easy data transfer led to new tools and approaches in public health. As early as 1995 a working group at the Centers for Disease Control and Prevention (CDC) had suggested an electronic system for public health monitoring and had developed a corresponding data standard. In 2000 a federal funding plan allowed the installation of technical systems based on this standard in several states (National Electronic Disease Surveillance System Working Group 2001). This introduced the National Electronic Disease Surveillance System to the United States that provided the infrastructural basis for a new approach to public health monitoring.

At first, this system was used only to facilitate data exchange between medical facilities, clinics, and health departments. Some health departments had already begun to test the use of nonspecific, nondiagnostic information to estimate the health situation of the population (Heffernan et al. 2004; Wagner et al. 2004). In addition, a federal initiative tried this approach to public health monitoring during high security alert events. For example, such a "drop-in syndromic surveillance" system relying on fax communication was established during a meeting of the World Trade Organization in Seattle in 1999 and during the inauguration of the American president in 2001 (Henning and Hamburg 2003). A permanent installation providing this monitoring approach emerged shortly after with the Electronic Surveillance System for the Early Notification of Community-Based Epidemics (ESSENCE) as cooperation between the Walter Reed National Military Medical Center and the health departments of Virginia, Maryland, and Washington, DC (Lombardo, Burkom, and Pavlin 2004). In subsequent years the ESSENCE system was copied to other states, parallel systems were introduced, and the CDC propagated its own nationwide syndromic surveillance system. These predecessors provided technical infrastructure, data formats, and experience to eventually implement this kind of monitoring as a permanent addition to public health monitoring throughout the United States.

Today the Biosense 2.0 system, employed and maintained federally by the CDC, is the integrated application of syndromic surveillance in the United States and has been installed across the nation. About four thousand hospitals enter their chief complaint data into this platform daily. This means that about half of the emergency department visits in the United States are recorded in this system (NSSP 2019).

The establishment of this additional tool in public health monitoring took some time and occurred against several kinds of resistance. Beside technical issues, questions of data security, and a struggle over respective areas of authority between state and federal institutions, many epidemiologists and public health practitioners have shown strong reluctance from the beginning. "If syndromic surveillance is the answer, what is the question?," as Arthur Reingold (2003) prominently asked. Doubts regarding the utility of this kind of monitoring were frequently articulated. Significant tension has been evident since 2001 between an ongoing political effort to strengthen and broaden the application and scope of this tool and its rather indifferent reception among reluctant professional users.

These tensions, hurdles, and the slow process of establishment are especially telling. The specific promises have apparently drawn political interest and support, even against the informed skepticism of professional users. The tool is apparently carrying conflicted paradigms beyond its mere functional quality. As such it is indicative of underlying shifts in engagement with public health.

If syndromic surveillance is the answer, global health security seems to be the underlying political question. The political interest in this new monitoring tool reflects an ongoing shift in the way public health is conceived as a political problem. The answer of syndromic surveillance becomes compelling given a specific globalization of public health. Several debates since the 1990s have cast the problem of public health explicitly in relation to the scale of the global. The term *international health*, where health policies are mainly about controlling the influx into the nation-state understood as bordered containers, was increasingly replaced by the term *global health* at that time. "We are on the verge of a 'global health village'" (Yach and Bettcher 1998, 736) stated a paper in the *American Journal of Public Health*. Importantly, this acknowledgment of global interdependencies has been formulated less as a reason for shared responsibility and first and foremost as a threat discourse. The "global health village" is predominantly depicted as a common breeding ground for newly emerging infections. Joshua Lederberg, Robert Shope, and Stanley Oaks (1992), for example, write that "in the context of infectious diseases, there is nowhere in the world from which we are remote and no one from whom we are disconnected" (v).

Several aspects of increasing global exchange and interdependence at that time, including travel, exchange of goods, and industrialized food production, are noted for their contributions to a rapidly changing microbiome (Armelagos, Barnes, and Lin 1996). An "emerging disease world view" (King 2002, 767) thus entails a turn to a specific epidemic rendering of ecology on a global scale: the conception of global interdependencies as decidedly pathogenic, unstable, and in need of constant monitoring.

However, syndromic surveillance is the appropriate answer, especially if the question of public health is viewed primarily as an issue of homeland security. The attacks of September 11 generally mark a far-reaching rupture in the US security discourse with particular implications for the politics of health. The public disclosure of the existence of letters laced with anthrax just a week later only raised the alarm further. Asymmetrical threats and the need for homeland security were regarded as the new normal, and anxieties about emerging infectious diseases were raised to the realm of national security (Keränen 2011, 467). This also put the threat of bioterrorism high on the political agenda, with the Bush administration implementing a national defense strategy against biological threats for the first time in US history (Cooper 2006, 113). Collective health status now surfaced as a highly plausible and vulnerable target, and the only viable solution was an early and rapid response. As Melinda Cooper argues, the security strategy after 9/11 shifted toward securing "life itself" in a broad sense. "Under the banner of the new intelligence agenda, certain defense theorists . . . were arguing that the scope of security should be extended beyond the conventional military sphere to include life itself" (Cooper 2006, 118).

Biosecurity became a goal articulated in the new security strategy after 9/11. Among several initiatives, the Defense Advanced Research Projects Agency (DARPA), housed at the Department of Defense, had been advised to develop an encompassing surveillance project as a response to the attacks. Its Total Information Awareness (TIA) project in the early 2000s tried to establish far-reaching tracing and tracking capabilities using personal data (Electronic Privacy Information Center 2004). One component of this system was intended to monitor signs of infectious diseases and biohazards. This Bio-event Advanced Leading Indicator Recognition Technology (Bio-ALIRT) signified an important push for early prototypes of syndromic surveillance. DARPA expertise helped to refine the automatic pattern recognition algorithms in the syndromic surveillance systems used today.

Syndromic surveillance as an innovation in public health monitoring and its enforced implementation after 9/11 in the United States reflects the interconnection and mutual support of two powerful discourses. First, the emerging diseases worldview offers a specific pathogenic perspective on the growing connections on a global scale, and second, the securitization of health positions public health as an issue of national security. Given both the inescapable reach of global interconnection and the instability and uncontrollability of the global natural and social environments, a constant and near-real-time situational awareness seems to be the appropriate political answer. Both the medical threat of emerging diseases and the security threat of bioterrorism are unknown and unknowable until they both express themselves through a sudden eruption in the regular public health situation (Donahue 2011). "The importance of strengthening public health surveillance has been the primary recommendation of all expert studies over the past two decades" (Morse 2012, 7).

TECHNOPOLITICS OF SYNDROMIC SURVEILLANCE

After reflecting on the question that is answered through syndromic surveillance, the central contribution of this chapter is to consider the relation of problem and tool in the other direction. The premise is to approach this innovation in public health monitoring not just as an expression of a certain problematization. Instead, I want to understand the performance of an established tool in shaping and reframing this understanding. How exactly is the problem addressed with the tool, and how does this questioning itself become politically performative? This line of reflection understands the example as a socio-technical arrangement and follows its concrete application in epidemiological practice on the ground.

Methodologically, syndromic surveillance is approached as a socio-technical infrastructure, drawing together technical installments, procedures, and practices. To include actual use cases and user experiences, three county health departments in Maryland, Virginia, and Washington, DC, were selected as case study sites. In 2012 qualitative interviews were conducted with representatives of public health administrations and with epidemiologists at the state and communal health departments regarding their day-to-day work with the ESSENCE syndromic surveillance system. The three-month case study period in Washington, DC, in 2012 further allowed access to briefings, webinars, and presentations. The case study covers experiences with a version of a syndromic surveillance system, ESSENCE II, that was established at that time. Further development has mostly refined the warning algorithm and allowed cloud-based data storage. The defining core functions and their uses were already present in 2012. After this initial case study period, the further establishment and maturing of the syndromic surveillance approach into today's National Syndromic Surveillance Program and the federal Biosense 2.0 system was followed through document analysis. The long process of innovation, deployment, and establishment of this monitoring approach in the United States since 2001 has produced a considerable amount of scientific reflection, gray literature, and conference proceedings.

Here I focus on how public health is conceived and remade into an environmental problem with and through the practices of syndromic surveillance. As noted above, the need for this tool is often legitimized by alluding to the instabilities of the global environment, for instance, as a reservoir of emerging diseases. This underlying conception of an epidemic ecology is specified in two directions through the practical demands of the system and its application.

First, syndromic surveillance helps to establish a shift toward engaging with public health threats. Prima facie, syndromic surveillance may be regarded as advertised: as a biosurveillance technology established to identify problematic events. In practice it is not the event, but the baseline needed for event detection that becomes the main object of concern. The system facilitates an environmental approach to public health with a shift from individual cases to situational awareness. Importantly, these situations are normalized through this lens and rendered as the unproblematic status quo to be maintained. Prevailing structural imbalances become part of the baseline and are no longer regarded as a concern.

Second, while the system broadens the perspective on the interrelations of humans, nonhumans, and the environment, the specific automated and data-based approach renders this understanding as mechanical cause–effect relations. Complex ecologies become factorized to tame their open, threatening potentials. This also implies a pathogenic conception and a turn away from broader, socio-epidemiological, and more structurally oriented health policies.

Assessing the Normal

As outlined above, the establishment of syndromic surveillance in the United States was very much dependent on the perceived threat of bioterrorism after 9/11. The acute need for an answer opened defense and military budgets to the notoriously financially deprived public health sector. Between 2002 and 2012 the US Congress spent US$ twelve billion on public health preparedness (Gursky and Bice 2012). Against this backdrop, the surveillance and early detection function dominated the narrative for a long time, presumably to help retain access to those budgets. Biosecurity was still dominant in US health debates in 2012 (see, e.g., Wielinga 2013). Interestingly, the users in my interviews had already shifted the narrative. They explicitly deprecated the early detection function for which syndromic surveillance was installed in the first place. "I can assure you that when there's something serious happening, an astute physician will be the first person to call us" (Representative of City Health Department, Washington, DC, audio of interview, Washington, DC, July 12, 2012, 00:51:23). As my informants repeatedly stressed, the system will never be sensitive enough to effectively detect the earliest signs of a potential hazard. Still, they appreciate other benefits of the tool. "It's the more slowly growing stuff that would probably more likely be detected" (Epidemiologist, Maryland State, audio of interview, July 2, 2012, Silver Spring, MD, 00:55:49). What the system does provide is an additional channel to observe the situation in general.

"Someone had already reported it [an outbreak]. And then now you have a list. So you can look at the system and see. I mean, are we seeing the same pattern in other areas? You know, so you can . . . it can help you. Where are they? You know. And then you can . . . then if it's dying out, you know, you will be able to manage it" (City Health Department representative interview, 00:09:33).

The ongoing stream of near-real-time data reveals a new perspective on the context of an event, providing an additional sensor for situational awareness. How can the significance of an event be assessed? Are there long-term trends and developments? Is there spatial clustering or other patterns? The warnings themselves are not taken too seriously, given the small number of problems of such monitoring. What the system allows instead is assessment and contrasting the relevance of these warnings against the broader context. The data collection itself is regarded as the main benefit, according to the users, much less the automated warnings of monitored irregularities.

The knowledge provided by this kind of monitoring is a "societal sense of what's actually happening," as I was told in another interview. "You know, the astute clinician might pick it up and might send up that red flag. And maybe even that red flag is earlier than what you're picking up with surveillance. It doesn't give you a societal sense of what's actually happening" (Epidemiologist, Virginia State, audio of interview, August 8, 2012, Baltimore, 00:06:28).

With the maturing of the system since 2012, this shift is now also officially promoted. In a recent newsletter, the CDC presents the New Jersey Health Department as a best-practice example in pioneering this shift with additional data sources. "New Jersey moved beyond using syndromic data for communicable disease into monitoring environmental and occupational exposures, opioid classifications, tick-related illness, and waterborne illness" (NSSP 2019, 5). Given its actual use, syndromic surveillance seems to be a tool less for preparedness than for situational awareness when confronting unstable socio-natural environments. Other best practices encompass the tracking of opiate consumption (Shekiro, Sussman, and Brown 2018), the effects of marijuana legalization, and social determinants of health such as domestic violence and homelessness (NSSP 2019, 8).

The collection of a broad range of partly nonspecific data allows epidemiologists to "complete the picture" (Maryland epidemiologist, interview, 00:42:35). "You can just click a button and you see with real time" (City Health Department representative interview, 00:24:04). A core benefit, according to many users' assessments, is the monitoring of the status quo rather than the automatically generated warnings.

If a relational ontology is assumed here, the interest is predominantly in how the practice of syndromic surveillance helps to establish its object, in this case public health, in certain ways. As Paul Veyne sums up: "Objects seem to determine our behavior, but our practice determines its own objects in the first place. Let us start, then, with that practice itself, so that the object to which it applies is what it is only in relation to that practice" (1997, 155). How, then, is the object of public health shaped through the practice of syndromic surveillance?

One defining aspect here is a shift from identifying the event to assessing the baseline of the normal. From the outset this surveillance approach was specifically conceived and installed to identify events of public health security. Syndromic surveillance provides ongoing monitoring of huge data streams. This makes automated pattern recognition indispensable. Pattern recognition needs an established baseline or horizon as a benchmark, so normalcy curves are crucial for the functioning of this kind of surveillance. Only with a baseline can the identification of worrisome spikes be delegated to automated routines of pattern recognition. This baseline eventually emerges as the main benefit. As sketched out above, spikes rarely happen, and users often do not trust them. What they gain from this system is an assessment of the normal state of affairs.

"One thing I look at every year when we have ice or snow, I look at heart attacks, people going out shoveling snow, a heart attack happens . . . or I put in falls, that is where they slip and fall" (Maryland epidemiologist interview, 00:41:15). The system allows the respondent to assess the normalcy of the state of affairs in winter. The goal here is not to detect the event of injuries or heart attacks, but to be assured about the routine state of affairs occurring again. Because the use of syndromic surveillance is about situational awareness, the contextualizing baseline of the regular state of affairs becomes the actual focus of this monitoring. My respondent does not ask the system "Are there heart attacks happening?," for instance, but "Do we see what is to be expected from the situation?" Through the ease with which several data streams are accessible with the click of a button, the baseline and its reverberations become the main object of scrutiny for professional users. The system incentivizes them to view public health as an environmental problem and look for the situation of an event rather than for the event as such. Are there weather conditions such as heat or snow, are there spatial clusters? Do the signals indicate contextual shifts?

Syndromic surveillance thus facilitates an environmental or contextual approach to collective health: environmental influences, social milieus, and spatial clusters are of special concern. But the interest in the influence of contexts is selective. As Carlos Frade remarks, such broad data-based approaches of pattern detection entail a flawed ontological conception of bias and structural conditions. "It privileges change (or, rather, a certain kind of change) while it is practically oblivious to the structural aspects which do not change but are inseparable from, and perhaps more significant than, what changes" (Frade 2016, 866). This privileging of change is integral to syndromic surveillance. The users neither care about the rare spikes, which are often false positives, nor about the injuries and events as such, which are the normal state of affairs; they are mainly interested in movements in the baseline. Consequently, structural conditions and existing health inequalities tend to be hidden by the normalcy of the baseline.

This monitoring renders the problem of public health as predominantly contextual and thus propagates a multifactorial etiology. But because this monitoring system uses individual emergency room visits and chief complaints as its primary data points, it remains unavoidably limited to agent-based modeling and thus constrained to the perspective of biomedical individualism. The structural and relational emergence of healthy and unhealthy bodies becomes a black box in turn that cannot be investigated. "In this view, disease in populations is reduced to a question of disease in individuals, which in turn is reduced to a question of biological malfunctioning. This biologic substrate, divorced from its social context, thus becomes the optimal locale for interventions, which chiefly are medical in nature" (Krieger 1994, 892).

Factorizing One Health

The practice of syndromic surveillance exerts a shift from event to context on public health thinking. A similar shift can be identified in the human–nature relationship. As a structured set of practices, syndromic surveillance articulates the problem of disease ecology in a specific way. The reason for adding syndromic surveillance to the tool kit of health monitoring stems partly from the growing perception of a troubled and permeable human–nature interface. However, practicing surveillance through these systems renders this perception in a specific way by factorizing dependencies and significantly reducing the meaning of ecological interrelatedness.

The debate on One Health in global health policy and the role of syndromic surveillance in this debate indicates this second technopolitical effect. At the core of this debate is the call for a more integrative perspective in approaching health threats. The rise of an emerging diseases worldview in global health and the growing acknowledgment of cross-species interactions such as zoonotic infections and antibiotic resistance in medical and biological science have both contributed to this argument. These rising concerns have driven a concerted effort, explicitly propagated as One Health in the Manhattan Principles in 2004 and heralded since at several conferences, through initiatives, and in medical journals (Gruetzmacher et al. 2021). A recent Lancet Commission report programmatically explains the core reasoning: "Thus, in our quest to ensure the health and continued existence of humanity, consideration must be given to the complex interconnectedness and interdependence of all living species and the environment—the concept of One Health" (Amuasi et al. 2020, 1469–70).

One Health has become a broad denominator for health policies trying to connect the often siloed approaches in medical science, public health, veterinary medicine, and ecological disciplines. "[One Health] goes beyond comparative medicine to link animal and human health together with the ecosystems in which they live, focusing on the current and potential movements of zoonotic diseases among human, domestic animal and wildlife populations and recognising that human, animal and ecosystem health are inextricably linked" (Welburn 2011, 614). Critics have observed a mismatch between such programmatic claims and actual health science and practice on the ground. In practice, public health, veterinary science, and global health are still mostly separated policy fields, each with rigid organizational, legal, and administrative barriers. Substantial and effective institutional changes that diminish such divisions are still rare (Stärk et al. 2015).

This is where practical applications bridging these silos are of special importance. Given the number of structural hurdles, syndromic surveillance is now increasingly presented as an existing practical application of the One Health agenda (Amato et al. 2020; Bhatia 2020). The very broad set of data gathered for syndromic surveillance, often encompassing chief complaints gathered in hospitals, weather data, animal shelter data, and data from veterinarian clinics, seems to provide the de-siloed approach necessary to the One Health paradigm of human–nature interrelation and interdependence. An EU expert commission on steps "towards One Health preparedness" accordingly recommends the strengthening of syndromic surveillance as one key measure (ECDC 2018, 11). The planned Triple-S syndromic surveillance is heralded as an important step toward employing One Health thinking in EU Health policies (Asokan, Kasimanickam, and Asokan 2013).

Becoming practical in a socio-technical setting also means articulating the premise of One Health in a specific way. The influence of environments on the epidemiological triad of environment–host–agent has recently been stressed in scientific accounts describing health-related phenomena and in policy debates (Corvalán et al. 2005; Parkes 2011). Claims for a geography of health, for example, strive to broaden the narrow perspective on pathogens and diffusion dominating recent accounts of medical geography (Kearns and Moon 2002; see also Prior et al. 2018). A settings approach that emphasizes socio-structural conditions is being propagated in current public health promotion efforts (Whitelaw et al. 2001).

Although relations and environmental determinants of health are currently increasingly acknowledged, what this actually entails is conceptualized quite differently according to competing underlying paradigms. It is in this struggle over underlying ontological assumptions that syndromic surveillance becomes technopolitically performative.

The point here is the mismatch between the relational entanglements emphasized in the One Health discourse and disentangling, factorizing, and categorizing surveillance practice. Approaching collective health as something to be translated into the stream of near-real-time data is eventually based on particular categorizations. For example, the chief complaints collected at hospitals need to be categorized into one of eleven syndrome categories (e.g., Botulism-like; Hemorrhagic Illness; Lymphadenitis) for the automated pattern recognition to function. "Categorization of clinical symptoms into disease syndromes is the cornerstone of syndromic surveillance" (Henning and Hamburg 2003, 291), but these categories still reflect the biomedical premise of a monadic organism. The data model is "built upon a binary thinking which creates hierarchies and boundaries between humans and non-humans, beings and the environment, diseased and healthy bodies" (Davis and Sharp 2020, 3).

In contrast, a substantial view of the premise of One Health demands the relational ontology of becoming entailed. As Gísli Parsson claims, "If the divide between nature and culture is to be bridged, it will be necessary to develop a science which takes becoming as basic . . . and conceives beings as islands of stability within the flux of becoming" (Pálsson 2013, 24). If the biosocial premise is taken seriously, relations need to be conceived as constituting their objects, rather than as forces and influences between fixed entities.

The categories used in the practices and technologies of current health surveillance actually rely on a pathological and biomedical rendering of health. Despite much talk of an integrated approach to health and the environment, these methodological requirements deny the biosocial understanding that would be required. Nicole Shukin describes this dilemma in her engagement with the current modalities of human and animal health: "While [One Health] is suggestive of a radical ontological breakdown of species distinctions and distance under present conditions of global capitalism, it also brings into view new discourses and technologies seeking to secure human health through the segregation of human and animal life and finding in the specter of pandemic a universal rationale for institutionalizing speciesism on a hitherto unprecedented scale" (Shukin 2009, 183–84).

Fundamental questions of species boundaries, the conception of health, and our relation with the environment are implicitly evoked in the One Health debate. Through syndromic surveillance practice, a certain technopolitical answer is given: pigeon-holing, quantitative measuring, identifying thresholds, and locating clusters. The understanding of health established within this infrastructure aligns with recent claims about an integrative perspective but renders this idea as a mechanistic biomedical model of calculable cause–effect relations. "One Health tends to remain a science based, clinically oriented, expert driven effort, with scientific disciplinary silos that do not connect or communicate" (Leeuw 2020, 28). The technical solution heralded to bridge these silos is in fact enforcing separation through the methodological paradigm it entails.

The Global Health debate since the early 2000s has emphasized the permeability of borders between both nations and species as a core public health threat. This invites consideration of the multiple assemblages of human and nonhuman entities comprising health as an effect of entanglements. Although the One Health claim programmatically follows this invitation, this paradigmatic shift cannot be found in actual policy delivery. One reason might be the kind of technopolitical answers currently given to the problematization of One Health. The automated, syndromic-oriented surveillance efforts sorting through nonspecific data streams entail a certain epistemology that tames the complexity of emergence into measurable determinants. Health is the absence of detrimental influences here, and the process of mutually conjoining human and nonhuman elements is effectively dismissed. Instead, the technopolitics of syndromic surveillance treat health simply as biology by measuring determinants, metabolic processes, and detrimental environmental factors. This conception of health "in terms of physio-psychological performance captures but a fraction of the 'real experience' of bodies in their associations, struggles and ambitions" (Duff 2014, 175).

THE PERFORMANCE OF SURVEILLANCE

Assuming a growing spatial and environmental sensitivity in global public health debates, this chapter has engaged with one specific policy response answering this sensitivity. A specific innovation in public health monitoring is taken both as an expression of this shift and as an active, performative element in it. With the help of a Foucauldian view of health and disease as historically contingent problematizations and perspective inspired by Langdon Winner (1980) on the political performance of socio-technical artifacts, the chapter analyzed public health threat discourses and corresponding infrastructures to understand the actual effects of this environmental turn.

Considering public health in this journey through epidemic intelligence infrastructures allows a more nuanced perspective on how a renewed emphasis of environmental interdependence in health science is translated to public health policies. The acknowledgment of human–environment interdependence runs deep today and is also expressed in the widely accepted Anthropocene narrative. A simple nature–culture dichotomy has to be reconsidered, given our immense effect on the environment, other species, and planetary processes and of our dependence on them (Dürbeck 2018). "The apparent dominance of the human species comes with a huge responsibility. Thus, in our quest to ensure the health and continued existence of humanity, consideration must be given to the complex interconnectedness and interdependence of all living species and the environment—the concept of One Health" (Amuasi et al. 2020, 1469–70).

Such changing awareness and related programmatic claims alter and are reworked if manifested in concrete policies and technics. Syndromic surveillance has been considered here as a specific example. The technical system can be read as an attempt to confront the acknowledged growing complexity and instability of the human–environment relationship. A detailed account hints at two technopolitical effects of this tool. The acknowledgment of an entangled human–environmental relationship is processed into single, actionable events that can be identified through the constant monitoring of the normal situation. Syndromic surveillance promises to provide a political handle on the unknown unknowns of indeterminable threats. Competing answers to this complexity, for example, the more holistic or structural perspectives on human–environment relations propagated in critical socio-epidemiology, are sidelined. The logic inherent in measuring the status quo to isolate events draws funding and attention away from the inherent, salutogenic aspects of the collective health situation. Structural determinants that help to establish long-term immunocompetence (Hinchliffe et al. 2013) or that strengthen resilience against infections and pathogens remain hidden in this version of epidemic ecology, which is formulated with and through current technical tools of epidemic intelligence.

CHAPTER 7

THE DIAGNOSTIC MACHINE

KNOWING DISEASE AND INSECURITY IN LIVESTOCK

Steve Hinchliffe

Inflated livestock numbers (Bar-On, Phillips, and Milo 2018), displacement of wildlife, and expansion of human populations (Bartlett et al. 2022) have altered and distended the human-animal interface. Epizootic and zoonotic disease risks have been amplified, selective pressures on microbial evolution have changed, and the dynamics relating to the emergence, persistence and transmission of drug resistant infectious diseases have shifted. These changes have made disease management and the management of drug resistance key biosecurity and human-animal health concerns in the farming and food sectors. In this chapter I focus on the apparent promise of new forms of diagnostic tools that can improve knowledge of diseases and inform more appropriate forms of treatment. Using the rich tradition of social science work on classification and diagnoses, I argue that current shifts in livestock farming toward on-farm and data-intensive systems may, paradoxically, perpetuate the very threats they seek to allay. I deploy the term *diagnostic machine* to refer to the social, economic, and material relations that inform how diagnoses are organized. In this formulation, diagnosis is a style of knowing that is not only concerned with making disease present, it also helps to consolidate and perform a set of social, material, and living relations. Diagnosis is an engine, not a camera (Mackenzie 2008), and diagnostics are embedded within and contribute to a capitalization of food production and marketization of health devices. The contention is that despite their promissory security logic, diagnostics may do little to ameliorate the ecological and health insecurities associated with livestock systems.

DIAGNOSTIC TESTS AND HEALTH SECURITY

Rapidly assessing, at the point of contact, whether a person, nonhuman organism, or environment is carrying a disease, or about to become infectious, can save lives. In a world characterized by the continuous emergence of infectious and resistant microbes, as well as an ongoing emergency of emergence (Dillon and Reid 2009), improved diagnostic tests and the ability to test *anywhere* and at *any time* have become a key part of a health and biosecurity logic (Street and Kelly 2021). If evidence-based medicine elevated the laboratory as an obligatory reference site (Daly 2005), point of care diagnoses are attractive for a variety of reasons. The first of these is speed, potentially allowing for test results to inform disease management decisions in situ. Second, portability effectively extends the network of diagnostic capacity, making tests particularly attractive in remote settings and in areas with few laboratories or trained staff to perform sampling and analytical procedures. The rollout of rapid test technologies to low-income countries, and to the 47 percent of the global population with little or no access to health care service–based diagnostics, is a key recommendation of a recent *Lancet* Commission on Diagnostics (Fleming et al. 2021). Third, the potential for high-volume testing means that results can, in principle at least, be aggregated and shared, making rapid tests particularly attractive for disease early warning systems, for surveillance, and for population health management. Fourth, tests ostensibly overcome problems of variation in clinical judgment, helping to standardize diagnostic and decision-making systems (Hanemaayer 2019). Fifth and related to this, the extension of diagnostic reach may be accompanied by a claim of improved precision and calculability. Harnessing the power of molecular diagnostics and bioinformatics has generated a promissory form of biomedical vision and pattern recognition, with enhanced data analysis and machine learning overcoming the limitations in human processing of results. For these and other reasons rapid tests have been targeted for accelerated development and emergency authorization during pandemics (Kelly, Lezaun, and Street 2022), and are subject to speculative investment across the life science sector. The result has seen a "proliferation of proprietary devices, effectively decoupling medical diagnosis from local infrastructures and expertise enabling diagnostics (and the data they produce) to circulate as commodities" (Kelly, Lezaun, and Street 2022, 191).

In the slow emergency associated with antimicrobial resistance, for instance, a rapid test panel with approval in 2021 from the US Food and Drug Administration (FDA) makes the promise explicit: "The Acuitas® AMR Gene Panel expands the diagnostic capability of clinicians to rapidly and simultaneously test for select drugs in 9 classes of antibiotics, . . . to aid in the identification of potentially antimicrobial-resistant organisms that might otherwise escape detection and hence can prevent prolonged inappropriate treatment of patients" (OpGen 2021). Making biochemical presences manifest provides the information needed to infer resistant disease threats and to specify effective treatment options. Claiming significant changes in disease knowledge and health outcomes is characteristic of an innovation ecosystem that needs to attract significant venture and private capital in order to make it to market (high-profile fraud cases and false promises in the diagnostics field, particularly in relation to the blood diagnostic tool Theranos, have produced a degree of wariness).

There is a variety of epistemological, material, social, and economic issues that affect product development and use, and that start to qualify some of the boosterish sales pitches. Kelly, Lezaun, and Street (2022), discussing accelerated development of rapid tests in response to Ebola, Zika, and COVID-19 outbreaks, highlight how devices may be poorly fitted to the dynamics of an outbreak and its management. Tests vary in terms of their specificity and sensitivity, standardized samples may be difficult to source, and in the case of a global pandemic, there have been gross inequities in terms of the distribution of and access to approved tests and to effective treatments. Trade-offs in terms of accuracy, speed, cost, reliability, utility, and ease of operation are common. During the COVID-19 pandemic, for example, in several countries there were debates regarding the utility of various testing devices. Despite their reduced specificity (or propensity for more false positives), antigen rapid diagnostic tests (lateral flow devices [LFDs]) provided advantages over polymerase chain reaction tests in terms of cost, portability, and availability of results at the point of testing. While potential inaccuracy limited their application in "test to protect" settings (protecting vulnerable populations), the ability to use LFDs at high volume and frequency made them ideal for population-based "test to release" processes as cases started to decline (Mina, Parker, and Larremore 2020; Larremore et al. 2021; Crozier et al. 2021). The example illustrates how epidemiological as well as social considerations shape the ways in which tests are calibrated and utilized. In animal health settings, the complexities are just as, if not more, notable. They include variations in testing environments, species, and breeds; diverging farming systems; the differences in monetary terms as well as the emotional value of animals; the cost, relative effectiveness, and availability of effective treatments; the regulatory environment; the availability of relevant veterinary expertise; the financial situation of farmers; the levels of financial compensation; and the option for "depopulation" in situations of heightened food systems or human health security.

As these opening comments might suggest, diagnostic tests are more than a matter of revealing things as they are, or laying things bare. They have a context, a health, disease as well as social situation. Where tests are being used, by whom, and for what purpose (for example to screen or to confirm), how they are accessed, and how and in what ways results are interpreted and utilized, all become part of a framework for their design, implementation, and evaluation. Given the current interest in diagnostic tests, including their development and rollout within fields as diverse as global health, public health, clinical practice, environmental epidemiology, and livestock and crop health, it is essential that these manifold associations are understood. To do so, it is useful to consider diagnosis and diagnostics in more detail, using the lens of social scientific and historical analyses.

"TELLING APART" AND "KNOWING AROUND"

Here are two indicative accounts of the process of diagnosing in the animal health (in this case poultry) sector.[1] They are both extracts from interviews with veterinary surgeons in the UK, and both refer to diagnostic results that relate to bacterial infections.

> We never treat in isolation, it's always in context. . . . [A] high level of *E. coli* doesn't mean that they need treatment. . . . [Researcher: What else is important?] How do the birds look on the farm, the mortality figures that have been reported that sort of thing. We have treatment thresholds that we go off.

> Sometimes you have to treat subclinical levels of enterococcus because it can [indicate] severe lameness later on. . . . [I]t's not only cheaper for the client, but you're actually using a lot less antibiotics, so it's much more responsible, rather than using it when they're two kilos heavier, and they're less likely to respond to treatment anyway.

Steve Hinchliffe

As these veterinarians see it, to treat or not to treat the birds is a matter that involves more than the indication or presence of an infection. Nor is it simply a matter of judging levels of infection. High bacterial levels may in fact lead to a decision to withhold use of antibiotics if other indicators are favorable, but likewise subclinical levels of a bacteria may require treatment in order to avoid complications. In this sense the vets perform a distinction between diagnostics and diagnosis. Their ability to place bacterial infections in context is key to developing an effective disposal or outcome.

Diagnosis is the science, and the art, of defining a problem. Its etymology is a clue to what is involved and provides two possibilities. In the first sense, the word stems from the Ancient Greek *dia-* (apart) and *gnosis* (telling or knowing), and thus "telling apart" (the meaning is clear in the sense of establishing or displaying diagnostic features), distinguishing one thing from another. Diagnosis is about identity, defining and recognizing disease signs and their associated causes. The onus is on sorting or spacing things out, allocating symptoms to diseases, and diseases to their etiology and to their appropriate treatment. Everything has its place. In the second etymological story, *dia-* can refer instead to "through" or "around" (examples of this preposition include diachronic and diameter)—so diagnosis becomes "knowing through," knowing "completely," or in less imperial mode, knowing around. If telling apart suggests a classificatory "spacing or sorting out" of disease types, knowing through suggests a different, temporal *and* spatial, register to knowing. It implies that knowing is achieved through working at it, experiencing something by living through it. This kind of knowledge is evident when an experienced clinician can read disease symptoms or signs through her ability to recognize and organize disparate clues. It's apparent in claims that patients may know their condition best of all. Or when a farmer's knowledge of their animals helps a veterinary surgeon develop a treatment plan. It also implies that knowing and experience are not things that can be reduced to a single facet. Instead, knowing around involves a process of drawing things together (Latour 1990), bringing disparate matters to bear on an issue or problem.

Telling things apart is clearly important: naming a condition may be vital in terms of implementing public health measures; isolating patients or farms; and accessing care, welfare, or insurance, and it is clearly important when the reason for diagnosis is to identify the most effective treatment. It is no good flooding a body with antibiotics if the disease relates to a purely viral infection. Distinguishing between infectious agents is organizationally and biologically useful. But knowing around is important too. Indeed, health may depend on it. While telling things apart involves an abstracted, spatial, or geometric separation of categories, knowing around involves more generous space-times of knowing, the need to piece together the evidence, and a requirement to be historically and geographically situated.

The tensions between these forms of knowing are certainly not new. The processes, labors, and consequences of telling things apart have been central concerns within the social sciences for over a century. Durkheim (1982 [1895]) emphasized the socially informed nature of classification and the consequent sedimentation of professional hierarchy and authority. Closer to the issue of disease, and possibly informed by (Durkheimian) social anthropologies of culture and medicine (Löwy 1988), the bacteriologist and clinician Ludwig Fleck developed a similarly social account of diagnostic practices (Fleck 1979 [1935]). Fleck emphasized the multifactorial nature of diseases (their failure to fit Cartesian coordinates) as well as the roles played by social settings and collective forms of knowing in generating diagnostic knowledge. As a clinician, Fleck understood the complex of knowledges that allowed a disease to be identified and understood. As a bacteriologist, he also appreciated the socially sanctioned nature of diagnostic tests (particularly the Wasserman test for syphilis), their development and specification, and in so doing suggested that the procedures leading to innovation in, and practices of, diagnoses were formed through a social collective that involved patients, doctors, and laboratory scientists, as well as a broader social sphere of interested parties (what he termed *eso-* and *exo-teric* or professional and lay circles) (Fleck 1979 [1935]). While his work was neglected for decades, it became formative for later sociologies of science, and notably for Kuhn's (2012 [1962]) account of scientific paradigms that emphasized the collective, routinized, and institutional bases for scientific practice.

The sociologist and clinician Marc Berg employed similar themes in distinguishing between cognitive and sociological approaches to medical problem solving. Cognitive approaches assumed historical information and medical examination results were regarded as facts that a physician needed only to lay bare, and treatment options were scientifically constituted, or fixed givens, against which obtained data could be matched. In practice, diagnosis was more likely to be sociological, wherein data were not so much "revealed" through examination than "reconstructed" in the process of consultation "so as to fit a certain transformation" (Berg 1992, 162). Clinicians were more likely to assemble accounts and to follow routines, institutional norms, and organizational procedures in developing a diagnosis that was defined by interrelations rather than biomedical facts alone. Sufficiency (in terms of good enough diagnoses and treatment) was determined by linking a problem to categories, *and* through the ability to realize economies of effort in time- and resource-pressured settings, to follow routines, and to adhere to institutional expectations. The point was not to lament how these social relations (workplace culture, institutional norms, and routines) infected or polluted the "real" business of proper diagnosis and subsequent treatment. It was to realize that good diagnoses and treatments took place through these social relations, not in spite of them. They were constitutive of what counted as good practice.

Foucault's *Birth of the Clinic* (1973 [1963]) and *The Order of Things* (1970 [1966]) provided a similar focus on the social establishment of disease knowledge, stressing the continual and somewhat undecidable struggles for order. Emphasizing the always incomplete nature of ordering, with its varied array of actors, agents, sites, and materials, science and technology studies, and actor network theory augmented this approach to the production of knowledge. Diagnosis, in this account, was conditional on social and material relations (Mol and Law 1994). Mol's (2002) research on atherosclerosis highlighted the work of assembling multiple and disparate practices (including laboratory tests, clinical examinations, patient accounts, and access to various forms of therapeutic practice) as a means to producing diagnoses. As Mol (2008) later emphasized, good medical care involved a process of assembling together these disparate knowledges and materials, an act of bricolage and tinkering, in ways that rarely made use of preexisting knowledge hierarchies (in this sense the laboratory report would not necessarily prevail over the clinical consultation and so on).

Some of these themes were picked up and utilized to lasting effect in Bowker and Star's (1999) *Sorting Things Out*. Drawing on Durkheim and Foucault, Bowker and Star analyzed several case studies, including the international disease classification system, as a means to reflect on the process of classification. Formally, classification assumes that a system of unique principles is in operation (a system that can be applied to new cases and produce coherent and predictable results, i.e., cases present themselves and are then sorted into specific types or instances of a condition), that categories are mutually exclusive, and that the system is complete. In this sense disease classification is oriented "toward a cause-and-effect [world] that resembles a set of slots, bins, or blanks on a form, even where it is multivalenced and multislotted: it is not, like disease and diagnosis, messy, leaky, liquid and textured with time" (Bowker and Star 1999, 172). Perhaps as a result of this mismatch between ontologies, classification tends to involve negotiation and compromise. Indeed, for a classification system to remain valid, there must be scope for flexibility. Bowker and Star noted that such systems tend to mix Aristotelian categories, where binary characteristics allow for a clear distinction between kinds, and prototypical classification, where a broader picture is implied and reasonable or metaphorical threads can tie a set of objects together (the frequent example is the class of objects called chairs which may have no strictly shared characteristics, but which are nevertheless recognizable as forming a category). One of Bowker and Star's key arguments was that most systems were in fact prototypical, but "they nevertheless have to appear Aristotelian to bear the bureaucratic burden that is put on them" (Bowker and Star 1999, 106). To work well, a system needs to be sufficiently flexible or aligned with the complexities of the world, but rules of interpretation and presentation need to sit on top of the classification as a means to smooth pathways and produce the semblance of a robust and widely adopted classification structure. These classifications are given form and consistency through agreement on standards, which exist to span communities of practice (they apply internationally or across a food production network, for example) and are often explicitly framed as legal or regulatory tools that become resistant to change.

This focus on the forms and practices of classification has generated a rich tradition in the sociology of diagnosis (Blaxter 1978; Jutel 2009; Jutel and Nettleton 2011; Jutel 2011), detailing the implications of diagnoses on the experiences of illness and the role of civil society in cases of contested knowledges. In animal and wildlife health the focus has also tended to be on practices (Hobson-West and Jutel 2020; Rock and Babinec 2008), on farm- and farmer-based expertise, (Bard et al. 2019; Mahon et al. 2021), interpretive work (Enticott 2012), and biosecurity outcomes (Hinchliffe and Lavau 2013).

This tradition in social scientific engagements with diagnoses underlines several key issues. First, diagnosis takes place in ways that are socially sanctioned. Second, in practice diagnosis is entangled with treatment and/or with the security options relating to culls, exclusion zones, trade relations, and financial considerations; it is a sociological as opposed to a purely cognitive process. Finally, categories and classifications hold or achieve stability if they are performed in ways that sanction discretion and flexibility (prototypical sorting). The implication is that diagnoses, diagnostics, and tests are part of an assemblage of matters. Understanding diagnosis requires us to account for this broader strategic apparatus and so to employ an ecological analytical framework. Just as important, we need to understand how diagnostic forms can play a role in consolidating or altering this apparatus.

"THE TYRANNY OF DIAGNOSIS"

In an interview, a UK poultry farmer (raising broilers) talked about the afterlife of a veterinary intervention that involved the use of diagnostics and treatments: "[Following a veterinary intervention] a couple of days later I'll get the post-mortem report. It will come as an email and all the sensitivity reports [in terms of responses to antibiotics] and all that. They have a computer system where it's all logged for you, a thing called DigiFlock, and all the medications, everything that they do is all listed on that, which is great when you do your Red Tractor audit. You just show everything!" Diagnoses are not isolated events; they circulate and accumulate as records. They are also audited by assurance organizations (in this case Red Tractor, a British farm assurance scheme) who act on behalf of food retailers and consumers to promote and assure consumers of standards in terms of food production and animal welfare. The digital trace makes life easier for the farmer in terms of record keeping and of course helps to generate industry-wide surveillance, records, and a progressive establishment of industry norms (concerning for example the sector-wide expectation for a reduction in the use of antibiotic medicines).

The context for and generativity of diagnoses are perhaps most explicit in the historian of medicine Charles Rosenberg's (2002) paper, "The Tyranny of Diagnosis." Rosenberg charts uneven and gradual shifts, largely within nineteenth-century Europe, from symptom-related, fluid, labile, idiosyncratic, and prognosis-oriented conceptions of disease (where a disease might be one of several stages in the progression of a complaint) to a focus on the identification of disease mechanisms (Rosenberg 2002, 242). In the terms used earlier, the work of diagnosis shifted from a concern with patients and their likely prognoses (a "knowing around" and a "knowing before") to a focus on the specific agents of disease (and a "telling apart").

For Rosenberg diagnostic practice retained a ritualistic and communicative quality (linking doctor and patient through a ritual of disclosure), but the shift of focus from patient to mechanism generated a new substrate for those relations, one that elevated biology *and* bureaucracy. The nineteenth- and twentieth-century proliferation of chemical, imaging and cytological techniques, as well as an expansion and conflation of diagnoses and treatments (with definitional specificity often conveyed via responses to specific therapeutics), simultaneously fueled the categorization, standardization, and stabilization of diseases as mechanisms, and reinforced nascent bureaucratic procedures and relations (from construction of hospitals to health insurance, health records, and access to treatment). As a result, at the turn of the twentieth century "diseases were accumulating the flesh of circumstantiality, both biological and bureaucratic" (Rosenberg 2002, 246).

This combination of the biological and the bureaucratic was key, for this was not simply science-led identification of pure forms or biological mechanisms. Ideal and typical disease pictures were not defined by their Platonic power alone but rather "their ability to acquire social texture and circumstantiality, to structure and legitimate practice patterns, to shape institutional decisions and to determine treatment of particular patients" (Rosenberg 2002, 250). This point is important, for this is not a history of the triumph of science over medical care—indeed, Rosenburg argues that practitioners remained skeptical of the shift from treating patients to treating diseases throughout the twentieth century. But the *spectacle* of diagnosis, the structuring of disease narratives, the displacement of a collective, cumulative, and contingent process with a "discrete act taking place at a particular moment in time" (256), and the ability to make experience "machine readable" (257), fueled a shift in diagnostic practices. Its power for Rosenberg was in the willingness of physicians to employ the constraining and yet empowering categories of nosologies even as they understood their limitations (and effectively a professional consent for Aristotelian and prototypical classification). The tyranny in Rosenberg's title lay in diagnosis becoming sedimented within a contexture of biological, social, and bureaucratic relations. In this sense it is not diagnosis, or diagnostics per se, that is the issue. It is the often visible but seemingly intractable nature and powers of the diagnostic machine that demand attention. Indeed, we need to recognize that "disease categories provide meaning and a tool for managing elusive relationships that link individuals to collectives, assimilating incoherence and arbitrariness of experience to the larger system of institutions, relationships, and meanings in which we all exist as social beings" (257). In other words, and rather like the model of interpretive flexibility and institutional power that Bowker and Star emphasized, diagnoses do not get their power from categories and pure forms alone. It is the contexture, the definitional flexibilities, and the boundary objects that allow various participants in a diagnostic network to confer its utility and power. In this sense it is the space-times of diagnosis that matter—the folding together of mechanistic tests and bureaucracies—that produce a robust form of knowing and doing. It's the process and norms that are generated through the folding together of biologies, bureaucracies, institutions, and so on that become the object for interrogation and critical analysis. So a key question becomes: How are things being drawn together? Through which spatial formations are the messy practices of medical disposal being linked to the legal and other authorities of fixed categories? And what might be the effects?

DIAGNOSTIC MACHINES

Clearly the logics and practices of diagnosis vary across settings. The issues will be different in public and clinical health settings, in lower-income health care settings, in animal health, and in environmental epidemiology applications. They will alter as we shift from identifying asymptomatic transmission of infectious diseases to early diagnoses of noncommunicable diseases, and move from estimating antibiotic susceptibility to helping a patient manage their diabetes. The question of how the space-times of diagnostics are being configured is an empirical question, one that requires an engagement with the relevant conditions. In this final section there are some indicative observations that open up the kinds of questions and issues that arise. The purpose is not to question the value of wider access to diagnostic testing. It is to insist on considering the social, technical, and institutional conditions that shape how diagnoses are formatted and in turn how the new diagnostic machine formats animal and health practices.

The implication of social science engagements with clinical practice in both human and animal medicine suggests that diagnostic tests are likely to be part of, rather than a replacement for, the work of piecing together a sufficient transformation of a disease or complaint. Even when a diagnostic is standardized and meets certain criteria (gold standards), tests tend to play a complementary rather than definitive role in assembling a health care plan. Telling things apart is conditional on a form of assemblage, or knowing around. Developing diagnostic tests for antimicrobial resistance is a useful case. Feted as a key part of a strategy to improve stewardship of a diminishing resource of effective antimicrobial treatments (WHO 2015), rapid diagnostic tests must deal with a wide array of conditions. There are, for example, a suite of technical difficulties, from identifying pathogens (and handling often polymicrobial infections) to predicting antibiotic treatment failure or success through correct identification of resistance genes or markers. The latter may be inconclusive as predictors of actual resistance, so virulence factors may need to be determined. Likewise, many forms of antibiotic susceptibility testing (AST) need to be able to mimic in vivo conditions that can change the resistance dynamics relative to in vitro tests. The point here is that the demands on any test and the complexity of data interpretation mean that it is more likely that rapid tests will need to be integrated carefully into existing practices if they are to be used with confidence and in ways that are clinically as well as legally robust (Li, Yang, and Zhao 2017).

This integration into an ecology of practices (Stengers 2005) is also performative. It generates new configurations as it alters the context of test adoption. For example, in the livestock sector, there were similar stories from veterinarians and farmers concerning the need to integrate diagnostics into existing practices, but also a twist in terms of diagnostic landscape. Livestock farmers, veterinarians, food processors, and retailers in the UK have been encouraged, in light of antimicrobial resistance risks, to move away from empirical therapeutics (or using antibiotics to treat all manner of animal health conditions prior to confirmatory test results being available) (Buller et al. 2020). The injunction to reduce antibiotic use, along with a comparatively relaxed regulatory regimen that existed for trialing and marketing diagnostic tools in the animal health sector and the time- and cost-pressured nature of food production systems, seemed to provide a perfect opportunity for the development of rapid, pen-side, diagnostics (Chan et al. 2020). However, developers found that the market and innovation environment was not conducive to developing tests (Bruce et al. 2021). Veterinarians and farmers were largely unconvinced that the tests would change their diagnostic and treatment practices. While they welcomed the idea of faster turnarounds, their experience cast doubt on the ability of tests to provide unequivocal results or clear solutions to animal health problems.

Veterinarians integrated tests and their indeterminate results into a *diagnostic process.* The latter included all manner of considerations, from markets and livestock dynamics to farm history and the complicated nature of often intercurrent infections. (A viral respiratory disease may still warrant treatment with antibiotics if there was a judgment that there was risk of secondary bacterial infection, while the presence of an enteric bacteria might be inconclusive when it comes to treatment decisions, the latter depending on the prognosis of whether and how the infection would develop in the lifetime of the animal and flock.) In sum, a range of practical considerations as well as social and market-based drivers rendered the notion of a definitive diagnostic test less relevant.

The twist is that smart diagnostics were nevertheless taking off within livestock sectors, and particularly within poultry and pig production sectors, though in a significantly different form. In high-throughput and relatively low unit-value sectors, where animals are raised in large numbers and in batches, the diagnostic machine is oriented to the herd and the breeding pyramid rather than to individual animals. In these settings, animal medicine is largely oriented to a form of population medicine. Poultry houses, for example, are increasingly outfitted with environmental and animal health sensors, monitoring everything from carbon dioxide concentrations in the air to litter moisture to bird weights and water consumption. Live feeds to a farmer's computer or phone enable early warning alerts of any departure from norms. If these warrant further investigation, the flock is sampled (usually this means carrying out a postmortem on a sample of birds). The results of this accumulation of signs, or knowing around, may or may not lead to treatment (depending in part on the standards set by the processor or retailer). But conditions and issues are recorded, and in some cases, it is future generations of animals who are treated with an additional or adjusted vaccine, or an adaptation of processes (from altering ventilation of buildings to increased electrolytes in diets to altering breeding stock genetics). In this sense the diagnostic machine is oriented not so much to the sick animal, or even to the herd or flock, but to the production process, the breed, and its genetics. Animal health is thus enacted through a different kind of body.

Diagnostics are being fitted to a productivist and progress-oriented version of livestock farming. The power of molecular diagnostics and bioinformatics, noted at the outset of this chapter, is folded into a biopower of population management. As Rosenberg (2002) suggested, the power of diagnosis is not derived from the revelation of pure forms or mechanisms. The limitations of nosological approaches are all too apparent to those involved in care practices. Rather, power resides in the ways in which diagnostics assimilate "incoherence and arbitrariness of experience to the larger system of institutions, relationships and meanings" (Rosenberg 2002, 257). The consequence is that it is becoming more and more difficult to farm without this machinery—auditors, retailers, food processors demand the circumstantiality, to structure and legitimate practice patterns, to shape institutional decisions. This informational world of livestock farming may suggest a future of improved animal health, better welfare, and reduced cross-species threats. Reductions in per-unit treatments of antibiotics may provide the proof that is necessary. But the obverse is also the case. The diagnostic machine may help to fuel further capitalization, reduced diversity, and increased livestock numbers (effectively canceling per-unit reductions in medical use). As with other audit driven systems, the focus on key indicators and oversight (surveillance) tends to perpetuate other oversights in the form of industry-wide externalities. It is the false assurance of apparent control in a world of exploitation and extraction that gives reason for concern.

SECURING EPIDEMIC ECOLOGIES AND DIAGNOSES

Securing epidemic ecologies is, it seems, conditional upon greater volumes of disease knowledge. The promise of the biomolecular, bioinformatic, and data-intensive approach to diagnostics tends to abstract data, testing apparatuses and their circulation from the social and material relations that make diagnosis, as opposed to diagnostics, a possibility. While new knowledge may be welcome, there is also a task for humanities and social science scholars to ask how this knowledge formats the world, and what kinds of circumstances are folded into (and bracketed out from) this version of the world.

Bowker and Star (1999) argued that, in order to function, systems of classification depend on prototypical forms of categorization, but nevertheless tend to emphasize their Aristotelian credentials as a means to gather authority and secure standardized practice. The subsequent weight of social science critique has tended to focus on this artifice, this apparent though never quite systematic imposition of abstract spatial form on the duration and messiness of life. In this chapter the argument for a much broader social science critique has been developed by focusing on the productivity and multiplicity of diagnoses and diagnostics. In engaging with the current ascent of rapid diagnostics, it is not just the artifice or even the value of telling things apart that needs to be dissected. Medical practitioners know too well the incomplete and often misleading nature of information that is abstracted from the contexture and evolution of living processes. More than this, the argument here has been to insist on a consideration of the ways in which social, technical, and institutional orderings are bundled together in any emerging diagnostic machine. The job of the social sciences involves mapping the ways in which innovation in and applications of diagnostics entrain and accumulate the flesh of circumstantiality; it is to properly evaluate the role of diagnostics in the production of new social spaces and configurations of health, agriculture, and environmental epidemiology. In the animal livestock sector for example, the march of diagnostics may be nothing short of an acceleration rather than amelioration of ecological insecurities.

PART III

THINKING WITH AND RETHINKING ECOLOGIES OF DISEASE CONTROL

CHAPTER 8

CONTACTS AND CONTACTS OF CONTACTS

EBOLA RINGS, RESERVOIRS, AND FRONTIERS

Ann H. Kelly and Clare Herrick

In the autumn of 2022 Ebola was once again back in the news as an outbreak threatened to spiral out of control in Uganda. Sparked in a rural village and driven by the more rare and less contagious Sudan strain, the initial hope was that the Ugandan government's expertise in outbreak response would quickly bring this epidemic to a halt. Containment, however, proved elusive. Like many other Ebola flare-ups that have bedeviled the continent in the past decade, efforts to trace and stamp out transmission chains were undermined by shortages in medical supplies and highly fraught politics, amplified by a public thoroughly exhausted by COVID-19 lockdowns. As transmission reached the capital city of Kampala, the global health community geared up to confront what was fast becoming a routine emergency outbreak scenario. Driven by ecological destabilization and widening health disparities, Uganda's Ebola outbreak was not an exceptional event, but an artifact of ever-expanding pathogenic "hot spots" that threaten to engulf the African region (Brown and Kelly 2014; Mitmann 2022).

The specter of ever-more-frequent Ebola outbreaks notwithstanding, the global health community has a powerful tool kit at its disposal: monoclonal antivirals, accurate diagnostic platforms, and, perhaps most important, an R&D pipeline of promising and proven-effective Ebola vaccines, two of which have received EU and FDA regulatory approval and are fully licenced for commercial use (though only against one of the five identified Ebola strains). That biomedical leap forward was the legacy of the West African 2014–2016 epidemic, which saw medical countermeasures in preclinical stages, "on the shelf," or otherwise nonexistent, ushered through an accelerated process of research and development facilitated through unparalleled flows of international money. A prospect previously unthinkable, the implementation of large-scale, multisited clinical trials under outbreak conditions demonstrated the possibility of bringing the exigencies of emergency response into alignment with the interests of for-profit actors.

The successes of that experiment have led to a new era in global health that has seen deadly epidemics transformed into "research enabling" environments (WHO 2016; see also Brende et al. 2017; Henao-Restrepo et al. 2016; Kelly 2018). Straddling what Andrew Lakoff (2016) describes as the distinct humanitarian and biosecurity regimes of global health, this emerging accelerationist mode of R&D reached its apogee in the response to COVID-19. The combined magnitude of the public health emergency, its territorial span and trajectory, sparked an unprecedented investment in medical innovation, compressing product development timelines from a decade or longer to a matter of months. Though the breakneck pace of innovation has cut across all fields of health care, most striking is the development of prophylactic measures, which in under a year from when the SARS-CoV-2 virus was first sequenced, produced several viable vaccine candidates to be distributed to billions of people across the globe.

Despite its tremendous achievements, emergency R&D has hardly been a public health panacea, as vulnerabilities persist and become exacerbated by a radical unevenness in vaccine access that the scale of distribution tends to obscure (Jensen, Barry, and Kelly 2021). Beyond a reprise of the glaring disparities between the Global North and South, those inequities further reflect the normative, fiscal, and epistemic fault lines that run through global health innovation. As COVID-19 becomes endemic and vaccine rollout remains socio-spatially distorted, the trade-offs between scientific rigor and public good, humanitarian need and national primacy, that had initially enabled the flow of novel medical innovations into global health response are losing traction.

Returning to the development of the Ebola vaccine and its public health afterlives offers a critical entry point to explore, with the advantage of some hindsight through the inexorable challenges in accommodating the standards and practices of the pharmaceutical industry into the strategies of outbreak control. However, the availability of effective vaccines and their role in bringing the West African epidemic to a halt should not be taken as an indication of a public health consensus on their deployment and use. The increasing spate of outbreaks has led to the call for new tools and improved strategies for testing and deployment—approaches that challenge the conventional ways in which the risk of Ebola transmission is imagined in time and space. Indeed, heated debates continue as to who should be immunized during an outbreak, how containment efforts might be best leveraged to generate the necessary clinical evidence for licensure, and what model of outbreak preparedness can sustain vaccine market demand.

To clarify those tensions and shifting stakes, this chapter describes three distinct spatial ecologies of outbreak control that abide under the conditions of global health acceleration described above. Our analysis begins with *the ring*—a targeted approach to vaccination pioneered in nineteenth-century England that gained public health fame in the 1970s by helping to eradicate smallpox. A form of epidemiological triage, vaccine rings are comprised of those at greatest risk of infection, quelling outbreaks while simultaneously providing an elegant solution to resource constraints. The ring's public health pragmatism became a key asset during the West African Ebola epidemic R&D, which saw this concentric immunological logic experimentally operationalized to better accommodate the exigencies of outbreak control within the investigative needs of a clinical trial.

Our second section focuses on *the reservoir*—arguably the foundational spatial metaphor for modern-day epidemiology (Lynteris 2023; Vaughan 1991). Reservoirs refer to human and nonhuman populations in which pathogens naturally circulate, posing a latent hazard and the ever-emergent threat of an outbreak. In addition to the complex temporal horizons of contagion, the reservoir also speaks to unrealized pathogenic potential, which in the context of research can act as a "standing reserve," as it were, for experimental exploitation (Helliwell, Raman, and Morris 2021; Hinchliffe et al. 2016; Keck 2019). While purportedly a species of fruit bat, Ebola's natural reservoir remains elusive—a much sought-after "smoking gun" for disease ecologists and a catalyst for biosecurity-driven wildlife viral prospecting initiatives across the African continent (Lachenal 2015). In the past decade the epidemiological figuration of the reservoir has expanded to include humans who have been shown capable of harboring the virus after surviving infection. The reframing of infection risk from zoonotic spillover to sexual contact compounds the stigmatization to which Ebola survivors are routinely subjected. But the disease risk that survivors pose to the wider population from the perspective of vaccine developers is also a valuable resource, not only in terms of the immunological insights this cohort provides for research and development, but also as a justification for a more comprehensive (and thus more financially sustainable) prophylactic vaccine policy.

Our final section considers the notion of *a frontier*. Admittedly, the concept of the "frontier" carries heavy rhetorical freight, its use so commonplace in innovation discourse as to have achieved the status of a cliché. We believe, however, that the term retains some heuristic value for an analysis of how biomedical R&D is being configured under conditions of emergency. As a moral imperative for intervention and future horizons of investment, the frontier is, to borrow a formulation from Angela Mitropoulos (2012, 5), "the primal scene of capitalist accumulation," marked expansion, uncertainty, and appropriation; frontiers are "the not-yet bounded territory simultaneously figured as the prospect of new markets . . . and the multiplication of points of exchange." The efforts to fold public health emergencies into the operations of commercial R&D reconfigures "disease risk" as "demand signal" and pathogenic landscapes as potential global health markets. In the humanitarian frontier, the incertitude of control and contagion give way to emergency exceptionalism that, through relaxation of norms and standards of practice, seeks to create a unified terrain of humanitarian response and commercial innovation (Elbe, Roemer-Mahler, and Long 2015; c.f. Herrick and Brooks 2020). That levelling involves a biopolitics that has to think again about the spatial assumptions that underpin the geometry of disease outbreaks, where pathogens are thought to cross over into healthy lives as if a pure space can somehow exist in contrast to an impure, diseased space.

RINGS OF PROTECTION

For Richard Hatchett, the CEO of the Coalition for Epidemic Preparedness Innovations, the West African Ebola R&D response marked a watershed in global health preparedness. "I think it says how far the world has come," Hatchet commented, "Five years ago, we were still debating about the ethics of doing research during epidemics, and really vitriolic debates. I think WHO's leadership in pressing for the inclusion of research and development during an epidemic response has been a game-changer. The fact that the world was committed to finding a way to conducting a randomized control clinical trial in this unbelievably complicated, challenging, and dangerous environment speaks to a paradigm shift in how we think about counter measures."[1]

Ann H. Kelly and Clare Herrick

The importance of research for outbreak containment is largely indisputable. In the face of a novel biological threat, epidemiological analysis, operational trial-and-error, and case-by-case clinical experimentation are the best, and often the only, tools available. That said, generating evidence in a public health emergency has always been an inescapably fraught and politically perilous endeavor, pitting personal privacy against public health intelligence, national sovereignty against global preparedness, and the lives of future patients against those currently under care (e.g., Kohrt et al. 2019; see also WHO 2016). The ethical compromises associated with a randomized clinical trial (RCT), however, are of a different order. In contrast to the prototypical ad hoc investigations undertaken during medical emergencies, RCTs are conducted at a scale and duration that entail greater risks to more people, who are more unlikely to experience direct benefits. RCTs are a tool designed to meet the needs of the medical marketplace, providing a standardized, statistical framework to interpret the merits of new products against the biases of patients and doctors (Marks 2000). The ethical justifications for distributing treatments at random, blindly, and ultimately for profit, quickly become untenable in situations where the deficiencies in health care provision and vast differentials in wealth and power can compel patients to participate in trials against their own self-interest. The potential for exploitation in a public health emergency, where patients are desperate for access to care, almost goes without saying (e.g., London et al. 2018; Sariola et al. 2019).

Those clear moral hazards, however, were offset by the uproar that followed the use of experimental therapies for Ebola such as ZMapp on foreign (and white) humanitarians, while that "right to try" was denied to frontline African health care professionals (see O'Dempsey 2017; Farmer 2020). As the outbreak threatened to devastate the region, the WHO concluded that Ebola's high mortality rate and transmission intensity demanded a new approach to control. Not only was it deemed "ethical to offer unproven interventions with as yet unknown efficacy and adverse effect as potential treatment or prevention" but the committee deemed that "in the current exceptional circumstances of the West African Ebola Outbreak, researchers have a moral duty to evaluate these interventions (for treatment or prevention) in clinical trials that are of the best possible design" (WHO 2014).

What constituted the best possible design, however, generated, in Hatchett's words, "much vitriol."[2] While widely regarded as the gold standard for medical evidence, individually randomized, placebo-controlled RCTs were a no-go for Médecins Sans Frontières (MSF), who found the prospect of denying treatment to mortally ill patients or, in the case of vaccines, highly vulnerable populations, untenable and in contravention of their principles. The proponents of the standardized RCTs argued that experiments deviating from a rigorously blinded and randomized design would risk equivocal evidence. Considering the complexity of the investigative context and the comparative ignorance regarding Ebola's clinical pathogenesis, many argued that anything less than a gold standard design would produce inconclusive outcomes and "waste scarce intervention-related resources, making them profoundly unethical" (Lanini et al. 2015, 738).

Ultimately it was a trial protocol designed by a team at the WHO that managed to thread the needle between compassionate access and investigative integrity. With the support of academic and charity partners, the organization coordinated the Ebola ça Suffit trial to test the efficacy of Merck's rVSV-ZEBO-GP to contain community transmission in Guinea. Ebola ça Suffit used a "ring vaccination" strategy, a novel design that targeted the contacts, and contacts of contacts, of a confirmed Ebola case. This "ring" of high-risk individuals was then randomly allocated to receive a vaccine either immediately or after a period of three weeks. By recruiting those most likely to be exposed to infection, Ebola ça Suffit functioned as a containment measure, building buffer zones against budding transmission chains. Recruiting high-risk populations also increased the trial's statistical power, ensuring the significance of outcomes against an ever-shrinking pool of potential participants as the epidemic passed its peak and began to wane (Graham 2019). In late January 2016, ten months after the trial began, no cases were reported in the rings that had received immediate vaccination, indicating a protective efficacy of 100 percent (Henao-Restrepo et al. 2017).

The pragmatics of ring vaccination have an illustrious public health legacy. Vaccinating populations at greatest risk was a strategy first implemented against a smallpox outbreak in Leicester at the turn of the twentieth century and then implemented on a wide scale during the WHO's decadelong effort to eradicate the disease in the 1960s and 1970s. In addition to a more targeted means of controlling transmission, the approach provided an elegant solution to vaccine shortage and considerable budget constraints (Heymann 2004). Forgoing territorial, population-scale coverage for strategically distributed communal barriers, ring vaccination built up the prerequisite herd immunity to put an end to all naturally occurring cases of deadly disease that had been circulating in human populations for millennia. "A triumph of management not medicine," according to the director-general of the WHO, Halfdan Mahler, the eradication of smallpox stands as a testament to the public health powers of epidemiological intelligence where bringing outbreaks to a halt is more a matter of "shoe-leather" field analysis and clever modeling than the extent of financial resourcing or even the efficacy of a discrete medical innovation (see WHO 2012).

Following the success of Ebola ça Suffit, ring vaccination was adopted as the recommended approach to Ebola immunization during an outbreak, and quickly came into play in the Democratic Republic of the Congo (DRC), where the second largest outbreak was about to take hold. Raging from 2018 to 2020 amid intense intercommunal fighting and targeted attacks on health care facilities, the Kivu Ebola epidemic, was described at the time by WHO director general Tedros Adhanom Ghebreyesus as "one of the most complex health emergencies we have ever faced."[3]

According to the WHO's "expanded access" protocol, Merck's rVSV-ZEBOV-GP was distributed by MSF to the contacts, and the contacts of contacts, of confirmed cases. For while trial results from Guinea had been encouraging, more data were needed to satisfy the regulatory requirements for full licensure, and thus access to the vaccine remained classified as "compassionate," deployable only in the extraordinary circumstances of an outbreak. But unlike its experimental distribution during Ebola ça Suffit, the safety and protective efficacy of the vaccine was, at this point, fairly clear. This made equitable access a focus of humanitarian concern. As the situation in the DRC deteriorated, MSF balked at what it believed was the WHO's slavish adherence to the ring protocol. "It's like giving firefighters a bucket of water to put out a fire," MSF emergency coordinator Natalie Roberts commented, "but only allowing them to use one cup of water a day" (WHO 2021).[4]

It was undeniable that limiting vaccination to two rings—contacts, and contacts of contacts—would be insufficient to control the outbreak; however, the WHO remained reluctant about widespread use. During intense deliberations, Merck announced that its manufacturing facility in Germany had unexpectedly failed to meet quality standards and the projected number of doses that would be available for distribution shrank considerably. Vaccine supply constraints and the still yet unknown duration of rVSV-ZEBOV-GP's protection dictated the critical importance of a rationed approach, but the logics of that new rationale were hard to parse. Since the outbreak's onset the response had strained under considerable gaps in epidemiological knowledge, a consequence of highly mobile populations distrustful of public health authorities and routinely displaced by conflict.

According to the expanded access protocol, vaccination was offered by an independent WHO-coordinated team of Guinean and Nigerian field epidemiologists, who had been previously involved in the imple-

For rVSV-ZEBOV-GP

Contacts of contacts
188.979 (64%)

INDEX CASE
Lab confirmed
EVD case

Contact: lived in the same household

High risk contact: close contact with patient body or body fluids, linen, or clothes

household members of high risk contacts

Contacts
60.472 (20.4%)

extended family

neighbors

Contact: visited the symptomatic patient

HCWs/FLWS

Potential contacts
Persons who can potentially be involved in the tertiary generation of cases
(e.g. 3rd level of contacts)
Variable number but estimated 50-100 per case

Potential Contacts
46.976 (15.8%)*
*eligible since June 13, 2019

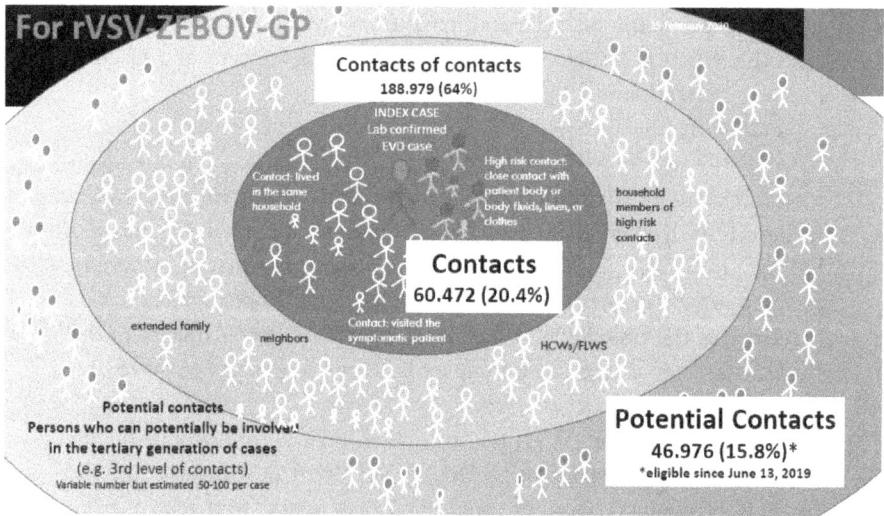

FIG. 8.1. Status of implementation of Strategic Advisory Group of Experts interim recommendations. Source: Graphic by Abdourahamane Diallo, WHO Health Emergencies Program, February 25, 2020. With the permission of WHO.

mentation of Ebola ça Suffit. While highly trained, the vaccinators' capacity to generate protective rings was only as good as the contact tracing that preceded them, work that was carried out as part of the national response by the Congolese. Delays in receiving the lists of confirmed cases and their (arguably inevitable) incompleteness, generated considerable ill will between the two teams, leading to accusations of corruption and cronyism. Compounded with increasing violence, much of which directly targeted health care centers, vaccinators felt hamstrung as their rings lost their epidemiological purchase.

A few strategies were pursued. First, vaccine teams established a series of strategically distributed and heavily secured sites for "pop-up vaccination" and largely abandoned following contact tracers into the villages. In addition, an additional third ring was suggested—the contacts of contacts of contacts—dramatically increasing the pool of those eligible for immunization. In principle, the third ring constituted an extension of, rather than a departure from, the "expanded-access" epidemiological rationale. However, in rural contexts, wherein the majority of inhabitants of a small village would be included, the additional ring essentially rendered the topology of containment isomorphic with a territorial approach. At the same time the WHO explored the possibilities of fractionating doses, a strategy that had worked in the region two years previously to quell an outbreak of yellow fever when vaccine stocks were

Ann H. Kelly and Clare Herrick

low (WHO, Department of Immunization, Vaccines and Biologicals 2016). The most diluted doses were offered to the "third ring," leaving those at higher risk of exposure receiving more potent doses.

Another obvious at-risk group was health care and frontline workers. Though often included in rings owing to their likelihood of exposure to cases, a more preemptive strategy seemed warranted both from an epidemiological and moral perspective. In addition to those in the immediate outbreak setting, health care and frontline workers in neighboring Uganda, Burundi, South Sudan, and Rwanda came into focus as compelling candidates for immunization. In the wake of the 2014–2016 West African outbreak, Ebola preparedness had become a regional priority, and the availability of vaccine candidates gave those preparations material substance. Given supply constraints, prophylactic vaccination with rVSV-ZEBOV-GP was not an option. But there were other vaccine candidates available, including Johnson & Johnson's (J&J) two-dose Ad26.ZEBOV/MVA-BN-Filo.

Giving one vaccine during an outbreak is challenging enough; locating the recipients after eight weeks to give them a second seemed almost impossible. J&J's preclinical work, however, was unequivocal: the booster dose elicited an exceptionally strong and long-lasting immune response. Unlike Merck's candidate, the two-dose Ad26.ZEBOV/MVA-BN-Filo was also multivalent, promising to protect against multiple strains of Ebola, not only its Zaire variant. What Ad26.ZEBOV/MVA-BN-Filo lacked was Phase 3 effectiveness data—evidence that was becoming hard to come by following the outcomes of Ebola ça Suffit. Considering its proven effectiveness, Meck's rVSV-ZEBOV-GP was the only appropriate candidate for compassionate access. But the outbreak in the DRC offered an epidemiological situation wherein the deployment of a vaccine at an earlier stage of development might be ethically tenable. Where transmission was extensive, populations mobile, and epidemiological risks largely unknown, it seemed to the WHO justified "to offer Ad26.ZEBOV and MVA-BN-Filo Ebola vaccines to those at some, but lower, risk of EVD in the context of an outbreak" (WHO 2021, 204).

Vaccinating health care and frontline workers in neighboring areas where outbreaks were deemed likely to spread, was not the same as a unilateral prophylactic campaign. "Those at some risk" constituted the emergency's *fringe*, a dynamic group that might eventually be included in a humanitarian response but not at its forefront (see Ranganathan and Balazs 2015). Critically, that liminal position at the edge of a potential outbreak gave them experimental value; rather than another large-scale trial among healthy volunteers, J&J could demonstrate their vaccine's effectiveness against naturally occurring disease. The unforeseen challenges to Merck's manufacturing capacities underscored the need to have other candidates, especially those effective against other Ebola strains. In issuing their recommendations, the WHO continually reiterated the need "to advance the clinical evaluation of other vaccines against EVD and to accrue additional information on their immunogenicity, safety and efficacy" (Were et al. 2019, 8), a commitment that is not easy to square with the availability of a preexisting effective vaccine. However, as a part of a preemptive (not strictly prophylactic) emergency response, vaccination of novel candidates could be implemented as part of a randomized trial of the kind justified in an emergency to halt transmission and protect the vulnerable. Under this delicately construed, strategically deferred, "just in time" containment, emergency vaccine R&D could feasibly continue.

This approach worked, not only in bringing the largest and "most complex" DRC outbreak to a halt but also in generating the necessary evidence for J&J's vaccine to achieve licensure. However, while it was a huge milestone in vaccine innovation, overcoming the problem of evidencing effectiveness did not solve the issue of ensuring vaccine availability. Without a viable market, J&J's incentives to invest in production, dedicating valuable resources and facility space, were limited—an issue of manufacturing priorities that would be tested during COVID.

There was another potential anchor for a vaccination strategy—Ebola survivors. The DRC outbreak resulted in 3,470 confirmed cases and 2,287 deaths—a mortality rate of 66 percent. Roughly seven months after the outbreak was declared over, another chain of transmission, again in Kivu, was identified as having originated among those fortunate 34 percent: an Ebola survivor persistently infected or experiencing a relapse. It is an epidemiological scenario that, in the history of known outbreaks of Ebola, was not only new, but threatened to become more common. Just a few months before the current outbreak in the DRC, a cluster of cases in Guinea was similarly traced back to a survivor who, despite having recovered from the disease over five years ago, still had a sufficient viral load to infect his partner via sexual contact.

The increased number of survivor-originated transmission events made the case to develop a protocol both for vaccinating those with latent infections as well as their close contacts. But unlike health care or frontline workers, the risks to which this group was exposed were harder to anticipate: the persistence of virus in survivors varies widely according to the individual and across time. Critically, that kind of targeted campaign raised serious ethical questions, risking entrenching the stigma survivors faced as a potential danger to their communities and undoing much of the work that had to go into reintegration and supporting their psychosocial needs (Delamou et al. 2017; James et al. 2019). But as

Ann H. Kelly and Clare Herrick

continuing Ebola outbreaks leave more survivors in their wake, the epidemiological—and commercial—significance of this group is becoming increasingly difficult to ignore.

THE PUTATIVE RESERVOIR

The concept of a natural disease reservoir emerged alongside the nineteenth-century bacterial revolution, though with the notable exception of Russian disease control programs, latent infections remained tangential to public health intervention and policy (Evans 2018; Honigsbaum 2016). The zoonotic origins of HIV and the securitization of health oriented to what Melinda Cooper (2006) identifies as pre-empting emergence placed new emphasis on latent infections, incubating future disasters and spreading without detection. Disease ecology radically expanded the scope of epidemiological analysis beyond the dynamics between pathogens and hosts, to examine the configurations of parasites, viruses, and bacteria through specific habitats, seasons, soils, and topography. A cornerstone of the One World One Health movement is appreciating that these pathogenic potentialities—the more silent zones of infectious disease—could be identified and eradicated before susceptible humans could trigger an epidemic.

However, integrative approaches to disease emergence do not necessarily do away with the metaphysics of power that characterize global health. Efforts to manage the ontological complexity of pathogenic circulations still operate across radically uneven political landscapes, and in their appeals to ecological holism, they can obviate and exacerbate preexisting vulnerabilities between species and among humans, producing a highly stratified vital politics (Craddock and Hinchliffe 2015). A ban on the hunting, sale, and consumption of wild animals (also referred to as bushmeat) was one of the first measures put into place following the declaration of an Ebola outbreak in West Africa. These measures were inspired by what was known about the cause of previous outbreaks, most often attributed to a single spillover event between a human and an infected animal. While the virus has been found circulating in a range of wild species, with bats believed to be the natural reservoir, at no point has an originating point of contact been identified. Perhaps more important is that once introduced into a human population, Ebola epidemics are driven by human-to-human transmission. Despite their ineffectiveness, legal restrictions on the sale and consumption of bushmeat were introduced across the region.

Local consequences of the bans—the markets closed, the fines levied, and the exacerbation of food insecurities—were profound (Bardosh, Leach, and Wilkinson 2016; Bonwitt et al. 2018). The ban's epidemiological irrationalities, however, found their logical expression in local outbreak narratives. The widely broadcast public health messages highlighting the risk of wild meat—risks that few had experienced firsthand—were read as evidence of the political machinations driving the supposed outbreak response. Wildlife protections have a highly charged legacy. Initially introduced by colonial powers, forest reserves were designed to protect "pristine nature" against the agricultural ravages of local populations—a hostile hierarchization of life and livelihoods that has persisted into latter-day biodiversity sanctuaries (Fairhead, Leach, and Scoones 2012). That a public health crisis could provide sufficient cover to appropriate land to the further impoverishment of local populations was hardly a stretch.

The "necropolitics" (Mbembe 2019) of the Ebola response became particular legible in media representations, the most iconic of which was in the August 2014 issue of *Newsweek*: "Smuggled Bushmeat Is Ebola's Back Door to America" (Flynn and Scutti 2014), featuring the image of a chimpanzee.[5] The grotesque sensationalism granted to "illegal African trade" over the ongoing plight of afflicted populations could not provide a more poignant illustration of what Achille Mbembe describes as the "forms of social existence in which vast populations are subjected to conditions of life [that confer] upon them the status of living dead" (Mbembe and Mitsch 2003, 1). The process of abjection continued throughout the response, as West African populations were simultaneously framed as both drivers of transmission and beyond the reach and reason of public health (Nuñes 2016; Benton 2017).

Since the end of the outbreak, however, zombification of African life has taken on a new biopolitical form, conjugated as survivorship. Nearly twenty-nine thousand people were infected during the 2013–2016 Ebola epidemic in West Africa, over half of whom recovered. The public health challenge this cohort presents is only just beginning to be understood. In addition to the often debilitating effects of post-Ebola syndrome and the lack of knowledge regarding the persistence of the virus in bodily fluids, this highly stigmatized group can face exclusion from the home, within the community, and the workplace (Ronse et al. 2018). Social and psychophysical vulnerability, however, does not exhaust the biosocial character of Ebola survivorship. *Survivors*, a term freighted with political purpose, have also served as a critical global health resource, volunteering as frontline health care workers, offering care for orphans, making public testimony in attendance to Ebola Treatment Units, and enrolling as research participants during and after the emergency response. Survivors were also recruited as experimental subjects, enrolled in clinical trials investigating the therapeutic potential of convalescent

Ann H. Kelly and Clare Herrick

blood and plasma. Their role in these public health and experimental endeavors in many ways constituted a means of recuperating a "spoiled" identity, leading in some cases to the formation of collective biosocial identities and political activism (Abramowitz 2017; Venables 2017).

Even the most sanguine accounts of Ebola survivorship, however, underscore the challenges of diagnostic uncertainty, an identity politics characterized by suspension and deferral closer to what Loclann Jain (2007) terms "living in prognosis." In the absence of any clarity on the long-term health implications of the disease, including the potential for relapse and reinfection, that "elegiac" existence is lived through the social and cultural repertoires through which Ebola survivorship is articulated. From their work with survivors in Liberia, MSF anthropologists Umberto Pellecchia and Emily Venables point to the problematic configuration of survivors as heroes: "Success stories framed around their status of being biologically alive, led MSF and other organisations to use them in the public arena without questioning the social foundations and political meanings of survivors' newly gained identity" (2020, 149).

The potential of survivors to act as a locus of imminent risk in future vaccination campaigns adds another temporal dimension to this biopolitical reductionism, one that folds the exigencies of humanitarian into the expectations of experimental citizenship. Calls have been made to track these newly discovered reservoirs, who some believe should, with their partners, be routinely vaccinated to keep the immune system primed and prevent them from transmitting. Others see their value in helping plug key gaps in what remains unknown about the durability of vaccines by providing an immunological model of the body's shifting antibody responses over time.

As the Ebola R&D landscape develops, we see how survivorship is further configured as a fungible resource, a demand signal by which companies like J&J can inform their development and potentially, depending on fair regulatory winds, orient their investment. At the same time the recent discovery of asymptomatic cases—a now-iconic pandemic category previously dismissed as epidemiological myth—has attenuated both the discrete biomedical value and disease risk that survivorship entails (see Löwy 2021). What remains clear, however, is the exceptional vulnerability of this "reservoir," whose health trajectory, and indeed onward survival, remains tenuously keyed to their articulation within a global health imaginary of emergency innovation and response.

REGULATORY FRONTIERS

In the wake of the West African Ebola outbreak, the WHO published its *WHO R&D Blueprint* for action to prevent epidemics, solidifying its role as an engine and architect for emergency global health research. The Blueprint lays out a strategic framework to prompt the "rapid activation of R&D activities" and establish a "research-enabling environment" under extraordinary times of emergency (WHO 2016, 19, 2). Among its core areas of concern are the identification of priority diseases, fostering collaboration between governmental, humanitarian, and commercial stakeholders, developing new investigative protocols and evidentiary standards for technologies in early development, and finally, overseeing regulatory and financing mechanisms for those activities. That agenda is driven by the new moral economy of pandemic preparedness, linking the acceleration of product development to the imperatives of infectious disease control. "While conventional surveillance, contact tracing and containment measures remain cornerstones of a health emergency response," the Blueprint states, "a repertoire of effective health technologies could be the key to pre-empting full-blown epidemics, and limiting their human, social and economic losses. The Ebola epidemic taught us that we can move faster to try to curb the spread of disease. By acting together based on a coordinated plan, we can accelerate the development of the vaccines, drugs, diagnostics and delivery systems needed to short-circuit emerging health threats" (WHO 2016, 2).

The Blueprint characterizes the belatedness of the Ebola response as a matter of technological deficiency—the crisis, in other words, could have been averted if key biomedical advances had been made. Emphasizing the lag time between disease containment and clinical product development, however, obscures the WHO's role in that delay, most notably the organization's failure to declare an official public health emergency for months despite clear signs that the outbreak was out of control (Garrett 2015). Postmortem analysis pointed to fundamental problems at the heart of the organization, from its dependence on the financial support and political will of national governments (Gostin and Friedman 2015), to the disjuncture between its normative purpose and operational capacities (Lakoff 2017).[6] But despite calls for its disbandment, the WHO's relevance was rewritten by what the Harvard–London School of Hygiene and Tropical Medicine independent panel described as its "key coordinating function in research and development R&D" (Moon et al. 2015, 2216). Pulled from the brink of "an existential crisis of confidence," the "WHO proved its capacity to lead, convene, coordinate, and establish norms among a broad range of public and private actors on research and development and data sharing" (WHO n.d. 2213).

That capacity was crystallized by the WHO's coordination of Ebola ca Suffit; situated at the threshold of humanitarian medicine and biosecurity preparedness, the field of emergency R&D opens up a new frontier of global health. This frontier is arguably a foundational metaphor for global health, an enterprise often preoccupied with territories and populations at the limits of administrative control and public health capacity (Farmer 2001; Giles-Vernick and Webb 2013; Street 2014). This imaginative register, inherited from colonial tropical medicine, combines spatial and epistemic dimensions: knowledge is extracted at the threshold of civilization and wilderness, from a zone of discovery that is perennially "emergent" and "vulnerable" (Brown 2011; King 2002). Propelled by a heady mix of humanitarian ideals, political aspirations, and commercial interests, the "imaginative project of the global health frontier," Brada argues, "legitimates a potentially endless horizon of intervention" (2011, 290).

The frontier has also been a foundational imagery in defining the relationship between scientific research and public welfare. *Science: The Endless Frontier*, a report prepared by Vannevar Bush, who headed the US Office of Science and Research Development during World War II, is the classic twentieth-century exemplar of this tradition. It exhorted the United States government to invest in basic "blue skies" research and articulated what later became known as the "social contract" between science, society, and the state in liberal democracies during the Cold War period. Bush's evocation of the frontier anchored this contract in America's foundational myth as an enterprise of constant expansion and betterment; it mapped a limitless horizon of progress driven by the engine of scientific research and attributed to the state the role of primary funder of science and of protector of its independence from immediate practical needs and private interests (Bush 2020 [1945]).

During public health emergencies, these territorial and scientific imaginaries of the frontier are fused and recombined. The epidemic frontier is simultaneously characterized by inconvenient borders marking the limits of human knowledge, or of governmental control, and the opening up of new routes for new scientific, commercial, and technological exploration. The emergence of novel pathogens is itself a feature of the ecological transgressions of globalization—the intensification and interpenetration of social, material, and biotic encounters that prompt infections to "spill" across species and spark wide-scale outbreaks (Nading 2013). Disease "hot spots" demarcate zones of unexpected encounters and unstable transformation, whether these are occasioned by environmental degradation or rapid urbanization or arise in an animal farm, antimicrobial-rich hospital, or along the forest's edge (Kelly, Keck, and Lynteris 2019; Mitman 2022). The unpredictability and anomalous nature of a novel outbreak, moreover, calls into question the forms of expert knowledge on which conventional disease control measures rely, requiring creative modes of thinking and cross-disciplinary collaboration, the kind of epistemic and ethical experimentation traditionally associated with "frontier research" (Lezaun 2010).

In Uganda that investigative dynamism has opened up new avenues for pharmaceutical investment. Because Merck's and J&J's now-licensed and -stockpiled vaccines are useless against Sudan Ebolavirus, WHO finds itself back to the drawing board working with governments and companies to integrate early-stage clinical trials into outbreak response. They have at their disposal an increasingly standardized tool kit—including ring trial designs, target product profiles, fractionated doses, and immunological extrapolations from animal models—to enable those early-stage vaccine candidates to make their way into arms, though the challenges of low epidemiological numbers and the politics of access to samples and cases are no less fraught. Initially two vaccine candidates, one from the Sabin Vaccine Institute and the other from the University of Oxford, led the field and were being readied for testing. That was until, to the surprise of Merck, one hundred thousand doses of an experimental vaccine against Ebola Sudan were found in the company's freezers in Pennsylvania. Merck had developed the product in parallel to its vaccine for Ebola Zaire that was used successfully in West Africa. But without the opportunity to test it in patients, the experimental vaccines had eventually reached the end of their shelf life and were set to be destroyed. Now it has been determined that they are still safe for use, and Merck plans to donate the vaccine to Uganda—which shares the technology of rVSV-ZEBOV-GP (now brand-named Ervebo) and is thus widely anticipated to be equally safe and effective. Considering the doses are soon to expire and with their experience working with the WHO, Merck's "wild card" donation is likely to take priority. This serendipitous discovery promises both a humanitarian and commercial payoff.

With the rise of emergency R&D as an authoritative way of framing the global health enterprise, it is thus possible to observe the formation in real time of new alignments between humanitarian, scientific, and commercial interests, marked by both political exceptionalism and technological utopianism. In these circumstances the humanitarian need that underpins the license to experiment operates across multiple timescales, spaces, and jurisdictions—war zones and industrial freezers—as evidenced by the sharp trade-offs between immediate humanitarian need and future benefit, political contestation over the sharing of biological materials, and rhetorical appeals to "global solidarity" against a backdrop of national protectionism and stockpiling. How Ebola vaccines conjugate these tensions illuminates structural characteristics of global health innovation, and the limits that an emergency R&D paradigm places on effective and equitable forms of international emergency response.

SPECULATIVE BIOSECURITIES

Medical advances have leveled Ebola—once the archetype of apocalyptic nightmares—into just one of many problems associated with African public health deprivation (Herrick 2019; Hirsch 2020; see also Livingston 2005). Indeed, comparing survival rates among humanitarian first responders, who were evacuated to receive care, and nationals in the early days of the outbreak, suggests that Ebola's high mortality rate is largely an artifact of what Farmer describes as "staff, space, stuff, and systems" (Farmer 2016). At the same time the discovery of the human reservoir has reframed Ebola in terms of chronicity—both in terms of latency of infection in individuals but also endemicity of the disease within populations and the chronic disruptions and crises periodic outbreaks can cause (Hubmann 2021; see also Reubi, Herrick, and Brown 2016; Whyte 2008). While still charismatic (Herrick 2019; Kelly 2018) in its capacities to ignite fear and mobilize action (categorized as a class A bioterrorist agent), Ebola's public health persistence has shifted the dynamics of contagion and containment, from speculative preparedness exercises and all-or-nothing securitized responses to an object of global health investment.

In this chapter we have explored the key set of spatial ecologies or epidemiological "infra-logics" attendant to Ebola's transition from exigent disaster to endemic crisis. These shifting spatializations of this highly charismatic disease imply a distinct imagination of what versions of outbreak control are possible and, in particular, the ways in which Ebola can be framed as a market opportunity and a unit of global health investment. The redeployment of the vaccine ring as an experimental protocol highlighted how risk of infection could be leveraged for the purposes of both disease control and of data generation. One could say that Ebola ça Suffit presents an inversion of the ring's public health value as a technique of economization, from protecting populations through targeted immunization to accelerating return on industry R&D investment. The framing of Ebola as a humanitarian concern is critical to managing that slippage, as the logics of the ring become distended and distorted in drawing the outbreak into the potentially-more-at-risk though less-than-safe fringe. It is here, within the projected populations of transmission—the countries that border an endemic region or individuals who live in proximity to a latent infection—that new forms of emergency research become possible.

We see a similar dynamic at play as the reservoir moves from a problem of ecological anticipation to one of public health management, from a purported natural host to a vulnerable human carrier. The degree to which the latter becomes an object of global health attention is keyed to the potential they offer to sanction new kinds of exceptional experimentation, commercial development, and discrete biomedical interventions. This emerging market presents what we have argued is a new frontier in global health, where pathogenic threats (speculative and real) become sites of marketization, containment, the evidentiary basis for accelerated licensure, and survivorship instrumentalized for future commodity demand. Along the frontier we see the routinization of emergency into a blueprint for action, new investigative protocols, norms, and standards that will rework exceptionality into a more stable set of humanitarian expectations around populations in need and future lives secured at all costs. It remains to be seen whether this accelerationist mode will be able to meet the challenges of contemporary pathogenic intensification and to what new forms of health security—and solidity—it will ultimately give rise.

CHAPTER 9

LAZILY INVASIVE?

CRITICAL MAPPINGS WITH AEDES MOSQUITOES, HUMANS, AND LANDSCAPES

Uli Beisel and Carsten Wergin

Transnational ties, sociocultural outlooks, technologies of transportation, knowledge, work, and lifestyle mobilities have a profound impact on global health. Not only humans and goods move; disease vectors and pathogens also utilize global connectivities to expand their habitat. As the SARS-CoV-2 pandemic shows, a better understanding of the mobility patterns of disease agents is crucial for early detection of outbreaks and successful containment (Adey et al. 2021). The question remains of how to come to terms with such expansionism of global health challenges. The same is true for mosquito-borne diseases. The geographical distribution of mosquitoes of the genus *Aedes* has continuously broadened over the last decades. And since *Aedes* have the vectoral capacity to transmit infectious diseases such as dengue, yellow fever, chinkungunya, West Nile virus, and Zika, their movement is associated with the translocation of disease threats to new tropical, subtropical, and temperate areas. Meanwhile, the question we want to tackle in this contribution is how the cartographies of mosquito distribution that are currently used can do better justice to the often fragmented and disparate appearances of these mosquitoes.

Aedes have populated different countries through human transportation, which in turn brings arboviral diseases to previously unaffected places (Lwande et al. 2020). *A. albopictus*, also known as the Asian tiger mosquito due to its striped legs and body, is considered to have originated in the tropical and subtropical areas of Southeast Asia, but is today found in many parts of the globe, including Australia, Africa, the Americas, and Europe. After being first detected in the United States via imported automobile tires from Asia, *A. albopictus* is now found widely in the Americas, where Zika and West Nile virus are a growing concern for medical entomologists and the local tourism industry (Sotomayor-Bonilla et al. 2021). In Europe five previously non-endemic *Aedes* species have been detected, inter alia *A. albopictus* and *A. aegypti*. Since 2010 several dengue and chinkungunya outbreaks in Southern Europe were autochthonously transmitted by *Aedes*. And the outbreaks are moving north: summer 2023 saw small autochthonous dengue outbreaks in northern Italy and Paris (Zatta et al. 2023).

In Germany the first records of *A. albopictus* were made at a parking lot along the A5 motorway, mainly used by tourists and freight vehicles coming from southern Europe, which demonstrated that living adult mosquitoes can be transported over extended distances (Becker et al. 2017). *Aedes*'s transgression of political and economic borders creates a significant challenge with regard to possible control mechanisms. Unwittingly aiding vector mobility, travelers and truck drivers become "companion species" of *Aedes* (Haraway 2005). Statistical mapping techniques have provided further evidence for human movement patterns that strongly assist the spread of *A. albopictus* in Europe and the United States after their introduction (e.g., Kraemer et al. 2019, 854). Here suitable breeding environments that are unintentionally provided by humans stretch all the way to random plastic covers used for pavement work in city streets. In Europe, further environments associated with habitat suitability include swamp areas, renaturation efforts, and flood control that are already known from historical attempts to mitigate malaria. Meanwhile, we contend that an individual plastic roof covering a construction site conveys a very different spatial distribution of *Aedes* than a cartography that spells out the geographical size of a swamp. So which of the two should mosquito and vector control focus on? How do we bring lived experiences and haphazardous practices together with the need for the rigorous monitoring and predictive requirements of models? And what are the methodological challenges and possible responses that a change of perspective, away from the territorial toward the fragmented, conjures up?

In this chapter, we start from the observation that *Aedes* mosquitoes get caught up in human transportation infrastructures, be they in trains, buses, or cargo transportation, and end up extending their biological habitat to areas where they were formerly not recorded. Our focus lies on how this expansionism is then processed to become an element of policy and decision making for mosquito and disease control. Since *Aedes* mosquitoes are biologically very adaptable and their eggs extremely sturdy, they make good "invasive species." While this is foremost a technical term used by entomologists and other biologists, it resonates differently if mosquitoes are labeled and monitored as such by organizations like the European Centre for Disease Prevention and Control (ECDC). Designating *Aedes* mosquitoes as "*Aedes* invasive mosquitoes" (AIM) feeds an imaginary of aggressive biological organisms who are expanding quickly and with a broad spatial range. This is visually substantiated by maps of linear progression (see fig. 9.1).

As we inquire into how one might complicate this idea of invasion and linear progression, we suggest it is necessary to include human infrastructures and practices into the maps of *Aedes* expansion, as well as surveillance and control strategies. The aim of this chapter is to explore how representations of mosquito distribution can be rethought to incorporate the lived experiences of mosquitoes and humans, and accordingly the nonlinear, the fragmented, and the complex. The hypothesis we put forward is that this becomes visible if we shift our focus away from technological control strategies toward the network ecology in which *Aedes* are embedded.

At the heart of this approach is an international research collaboration of entomologists and social scientists that we are involved in and that seeks to build close, interdisciplinary engagement in order to make an intervention into the health security debate. In the first half of our chapter, we make a conceptual case for why a focus on the network ecology enables us to systematically include the lived encounters between humans and mosquitoes into our maps and analysis. The second half will play with some methodological possibilities for doing so. In sum, our chapter is interested in the relation between mosquito and vector ecologies, knowledge production about these ecologies, and how we might alter such knowledge production through the introduction of critical cartographies that lead to more nuanced, geographically and praxiographically attuned control strategies.

How to best capture mosquito-borne diseases on a map is of course not a new question. The mapping of infectious disease vectors has a long tradition in the natural sciences. Maps and diagrams have played a crucial role in visualizing and understanding epidemics, as they allow one to plot biological, ecological, infrastructural, social, and cultural factors in space and time. Their origins are also closely intertwined with imperialism and colonialism (McNeill 2010; Beisel and Wergin 2021). They have consolidated power over definitions as well as control measures, as John Snow's famous map of London water pumps and cholera transmission and distribution illustrates (Monmonie 2011). Engelmann even suggests that spatial diagrams "come closest to *represent[ing]* the epidemic itself" (Engelmann 2019, 89, emphasis in original).

Today, with rapid environmental change and associated health risks, modeling disease threats and the dynamics of epidemics has become a crucial element of not only political decision making but also popular and media discourses. Yet while there is much to learn from both the explanatory as well as the performative power of maps and diagrams (see also Engelmann, Humphrey, and Lynteris 2019), social scientists have criticized their reliance on quantified and plotted data as privileging certain control strategies. For instance, Everts calls COVID-19 "the dashboard pandemic," suggesting that it engenders a territorial and militaristic approach to disease control at the expense of recognizing global entanglements and local care possibilities (Everts 2020). Along these lines, Hinchliffe, in an opinion piece early on in the COVID-19 pandemic, calls for more attention to the spatial complexities and heterogeneities of associated knowledges, encouraging us to "practice the art of knowledge triangulation" (Hinchliffe 2020, n.p.).

We argue that ecological entanglements become particularly salient if we follow the vectors through critical cartographies and related ethnographic engagement, and in doing so diversify knowledge production and potential control practices beyond the abstractions of modeling and dashboarding. Attending to spatial complexities is especially pertinent for vector-transmitted diseases, where the meeting of (at least) three organisms—host, vector, and pathogen—and their environments are necessary for disease transmission. Socio-ecological and infrastructural dimensions of disease control and transmission are crucial to understanding where disease hot spots occur, and how they align with various material proximities (Brown and Kelly 2014). Ultimately then, integrating heterogeneous knowledge is not aimed at generating a holistic model or explanatory system as maps and diagrams tend to suggest, but to deepen how lifeworlds and mobility patterns of human and nonhuman organisms come into contact.

Given the importance of understanding the patterns and drivers of mosquito distribution, what has been lacking so far are systematic interdisciplinary studies that aim to link conceptual innovations in the social sciences with cutting-edge studies in entomology and disease ecology, while also working closely with communities (Bartumeus et al. 2019). Over the last decades multispecies studies of the entanglements of humans, vectors, and pathogens have emerged as an approach that can contribute to such attempts (Adey et al. 2021). New methodologies, such as multispecies ethnography and an anthropology of epidemics, have been developed (Kirksey and Helmreich 2010; Kelly et al. 2019). They challenge conventional cartographies and categorizations into developed versus emerging regions through mobile documentation. We approach these entanglements of mosquitoes and humans through the concept of infrastructures and the experience of "infrastructuring environments" (Blok, Nakazora, and Winthereik 2016). They define *infrastructures* as the patterns that societies use to coordinate movement across the environment. This entails organizing, comprehending, and controlling a wide range of relationships. Humans are selective when building infrastructure around the world and favor some species over others when doing so. The transportation of livestock and other domesticated animals receives ample consideration in decision-making processes, while unwanted species and diseases travel along the same human-made infrastructures unintionally.

That said, environments are not passive recipients of infrastructures either, but have a direct effect on how they are created and put to use. Tsing and colleagues's *Feral Atlas* (2021, n.p.) is an example of patchy maps that bring together "invasion, empire, capital and acceleration" and critically investigate the connections between infrastructures, capital investments, and other ways of structuring not just human environments but also multispecies geographies. We suggest that spatial representations of mosquito distribution demand a similar, broader understanding and methodology of mapping in order to flesh out their complex socio-ecological and infrastructural factors. For instance, what biologists and ecologists have called the "colonization" of Europe by *Aedes* species suggests, "at least two independent *invasions*, one from a northern Asian lineage which colonized Albania and northern Italy (likely via passive dispersion from the USA), and another from a Southern Asia lineage, which colonized Greece and is present (in minor proportions) also in central Italy" (Pichler et al. 2019, 15, emphasis added). As infrastructures transform alongside those who inhabit them, critical cartographies that seek to account for the heterogeneous, spatially and socially complex character of *Aedes* distribution must move with them. Such representations need to "go-along," need to be fluid and in process, rather than be artificially put on hold. Ecology then does not simply include biological factors, but rather a network of human and more-than-human actors, as well as the underlying infrastructural means that cater to it and that a critical cartography of mosquito distribution and related disease vectors needs to account for.

For the remainder of this chapter we will thus not speak of an ecology as the network but use the concept of "transecology" as introduced by Carsten Wergin to emphasize the heterogeneity of networks, both in character and also in time and space (Wergin 2023a). The prefix *trans* reminds us that the ecology that in our case is conjured up by encounters between humans and mosquitoes, infrastructures and diverse means of transportation, is never fixed. Furthermore, we feel that it is the particular quality that the social sciences bring to the study of related global health challenges that allows us to keep policy open to account for this. In order to overcome the colonialist rhetoric outlined above, we think it is worthwhile to consider multispecies mobilities in transecological terms in order to enhance narrations and emphasize processing and aesthetics, first and foremost in cartographies, as momentary expressions of people's entanglements with local environments (Wergin 2023b). Taken together, this suggests to follow mosquitoes as they move with and in human infrastructures and developing new cartographies that generate knowledge about the knots in which our shared lifeworlds meet. With the aim of substantiating this argument, the following section concerns the spatial imaginaries and empirical realities of the changing *Aedes* distribution and habitat in southern Germany. We complicate the invasion map of the ECDC by zooming into the microhabitat of *Aedes* mosquitoes in one town in Germany. In the remainder of our chapter we will then introduce what we have termed the "infrastructural go-along" as a new ethnographic method aimed at elucidating the emerging transdisciplinary cartography of entangled mobilities of humans and *A. albopictus* mosquitoes in Germany.

AGGRESSIVE INVADER OR LAZY MOSQUITO?

While vector-borne diseases have persistently been present as lived realities in much of the Global South, they are a new and rising health challenge in the Global North. To detect disease outbreaks early and to develop successful, locally supported control strategies demands a better understanding of the entanglements of human and mosquito mobility. In other words, the unintended "invasion" biology and biography of *A. albopictus*, given that it is considered one of the most aggressive vector species of human diseases, has high geopolitical and economic stakes that demand a stronger involvement of social science and humanities researchers. Yet a systematic transdisciplinary collaboration, as the one pursued in the aforementioned project, that aligns research strands of human, vector, and viral mobility is still sparse. The International

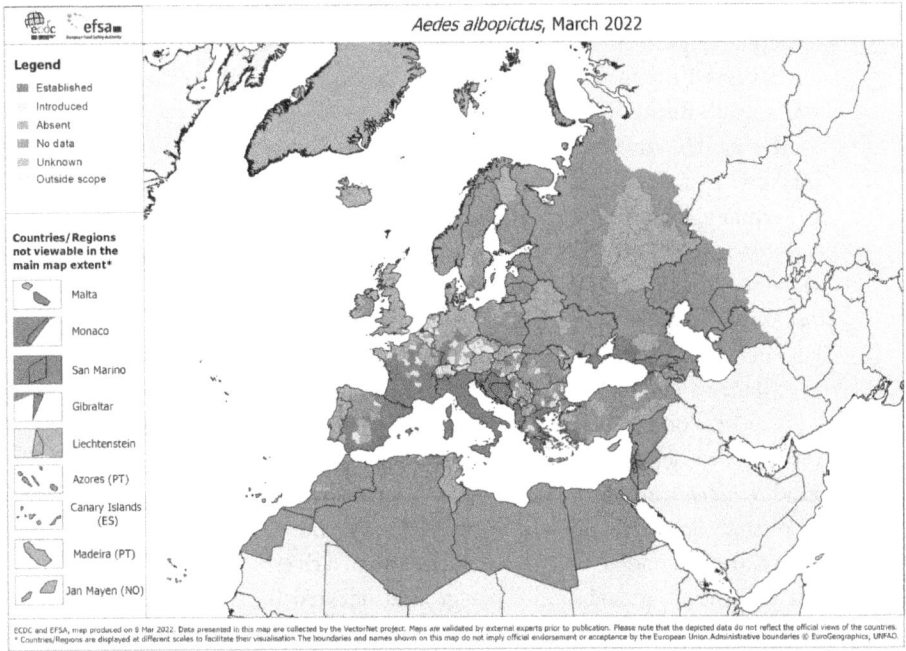

FIG. 9.1. Distribution of *A. albopictus* from southern to northern Europe. Source: ECDC and European Food Safety Authority (2022).

Health Regulations (WHO 2005) encourage surveillance at points of entry such as ports to monitor introduction pathways, causes, and routes of invasion and connect this to different biological and ecological aspects. More recently the movement of *A. albopictus* from southern climates to northern Europe has been documented in a map by the European Centre for Disease Prevention and Control (ECDC).

Based on these maps, it is easy to subscribe to a linear invasion narrative of *Aedes* mosquitoes, moving from tropical places to newly warming ones, from the Southern to the Northern Hemispheres, and from southern to northern Europe along major road corridors, such as the German highway A5 in southwest Germany. In this section we aim to complicate this picture a bit.

In the mid-2000s close to the first gas station after crossing the Swiss border, entomologist Norbert Becker and his team were the first to discover *A. albopictus* in Germany. They had been alerted by the insects' growing presence in northern Italy and Switzerland and did routine monitoring through ovitraps (Becker et al. 2012). In a personal conversation Becker explained to us that they assumed mosquitoes travel with tourists and trucks to find places in Germany with conditions amenable to their biological niche. Over the ensuing years Becker and his team thus followed the mosquitoes' "invasion" along the A5, from the Swiss border up to Becker's home region, the Rhine-Neckar valley.

The ECDC tracks this spread of *"Aedes* invasive mosquitoes" into Northern Europe. In its distribution map (see fig. 9.1), the southwest of Germany, from the border with Switzerland and (by now) up to the Netherlands, roughly following the Rhine River's path, is marked in bright red, meaning that *Aedes* mosquitoes have "established." But what does "established" mean? Which spatial pattern does "established," and also "invasion," refer to? And how is this pattern related to mosquito surveillance and control?

A closer look into the story of *A. albopictus* in one small southern German city helps address these questions. Becker and his team first found *A. albopictus* at a campground, consistent with the assumption that the mosquitoes hitch rides in tourist vehicles. The campground resembled a rather extended parking lot at the end of a riverside meadow, with volleyball fields, a café, and a playground, located in the heart of the picturesque southern German city of Heidelberg. The campground is surrounded by affluent residential areas, where old beautiful inner-city villas line the streets and the parking spots for campers at the riverside. For mosquitoes this is not a bad place to alight, as there is water, plenty of green space, and humans to bite. Thus for various unspecified reasons—maybe because it was the first place in the city where *Aedes* were detected, maybe because it is a prominent spot in the city center and close to more affluent areas—Becker and his team were provided with plenty of resources to successfully eliminate *Aedes* after just one season.

A few years later though two other areas became infested with *A. albopictus* and in these places the mosquitoes were able to overwinter. Of those two areas on the edges of the city, one is an allotment garden where people grow vegetables, plant flowers, and have small garden huts to spend afternoons and evenings outdoors. Interestingly, a second allotment area, a mere one kilometer away from the first and connected to it via fields and a road, has not been infested. Instead, the mosquitoes have been found some five kilometers farther in a neighborhood with detached single-family homes, an area that is significantly more urbanized, with tram stops and increased car traffic.

As opposed to the campground, it is not as obvious how the mosquitoes arrived in these two new areas. However, *A. albopictus* have come to stay. They have been present for a few years, and in a program funded by the municipality, Becker and his team no longer define their goal as eliminating but merely diminishing and controlling their population. To accomplish this, they position ovitraps in gardens to monitor the population's size, and regularly apply the larvicide BTI (*Bacillus thuringiensis israelensis*) to standing water, be they puddles or in containers. Furthermore, sterilized male *A. albopictus* are released during the spring and summer months to diminish the population. Residents are informed and sensitized to mitigations they can undertake via posters and door-to-door campaigns.

What spatial pattern of *Aedes* distribution does this southern German city display? First, the *Aedes* "invasion" remains fragmented. The mosquito has been detected only in very few parts of the city. While the ECDC map coated in red suggests an "invasion" that is encompassing and linearly progressing from south to north, our zooming in suggests a rather spotty pattern with small areas where mosquitoes are established and larger areas uninhabited by *A. albopictus*. Furthermore, we have discovered particular areas where *Aedes* have lived for several years, while in other places they could be eliminated by mosquito control activities in only one season. In sum, the pattern for the introduction of *Aedes* must be considered spatially and temporally complex and shifting. As the example of the two allotments close to each other shows, despite similar ecological conditions and a green corridor and car traffic between them, *Aedes* have made their home in only one, while the other remains uninfested (for now). As one entomologist on Becker's team put it: "*Aedes* are in the end rather lazy. If they are fed and watered, they are happy to stay put."

These first observations suggest that a fragmented, slightly random pattern might be more adequate to represent the distribution of *Aedes* in southern Germany. As the entomologists explained in a personal interview (June 2022), the mosquitoes are found in certain delimited and small areas all over southern Germany, mainly in allotments and other hot spots. The spatial logics of biological invasions are thus more complex than a blanket "invasion" suggests. This is of course stated in the ecological literature, yet when entomological knowledge is translated into policy, simplifications such as the ECDC map dominate both thought and practice. As a result, naming *Aedes* the "Aedes invasive mosquito" by the ECDC and a related scientific network perpetuates an imaginary of the mosquitoes as aggressive and on a warlike trajectory to "take over" areas and countries.

This short vignette prompts us to ask what difference it would make to characterize *Aedes* as "lazy" and opportunistic, rather than aggressive and invasive? Taking a look at the literature on spatial patterns and logics of mosquito distribution, Shaw and colleagues (2010) suggest that two spatial ontologies underlie malaria control activities: first, a "transcendent verticalism" that leads to aerial spraying of insecticides, as space is seen from above and understood through grids and linear partitions on a map. They distinguish this from "immanent horizontalism," which focuses attention on the painstaking identification of breeding grounds that is "intimately attuned to the spaces of the bug's life" (Shaw et al. 2010, 380).

This basic differentiation is further specified by Kelly and Lezaun (2013), who distinguish territorial and bionomic approaches. A territorial approach would focus on mapping and controlling all breeding spots or mosquitoes in a given region, which prevails in the discussion of the European *Aedes* "invasion," the mapping of the ECDC, and others. Bionomic approaches, on the other hand, try to understand an ecological network and formulate control activities accordingly. As a result, "[ecological] resources necessary for mosquito reproduction will be unevenly distributed across space, [as it is] the system of proximities and distances between humans, mosquitoes and pools of stagnant water, not the boundaries of irrelevant territorial jurisdictions that provide the most practical topology for a successful anti-malarial campaign" (Kelly and Lezaun 2013, 97). In a sense, Kelly and Lezaun's argument even goes beyond what they themselves argue: it is not just ecological networks that determine the mosquitoes' reproductive rate and spatial clusters, but a spatially thick, heterogeneous, transecological network of ecological, social, and infrastructural factors.

These heterogeneous and complex spatial features are highlighted by Nida Rehman: "The mosquito breeding ground represents the conjuncture of historical inequities, political ecologies, and material relationships not always amenable to the simplifications that systems desire. Disease surveillance infrastructures do not always 'see' like the mosquito" (Rehman 2022, 1452). Her observation offers a first point of entry for what we propose with the infrastructural go-along as a method through which to acknowledge the uneven power relations that still prevail in conventional representations of mosquitoes and disease dispersal. As social scientists follow the mosquito through diverse infrastructures they are to go along with the latter rather than interpret mosquito movements according to presupposed narratives of invasion and control. In this sense, we welcome the distinction between territorial and ecological space, all the while it remains important to further complicate this distinction through time. Thus our perception of space is transecological. Rehman shows that infrastructural neglect and inequalities are closely linked to mosquito prevalence and the likelihood of transmission, and that the territorial control methods in a technologized version of "real-time epidemic intelligence" (1469), applied to the city of Lahore in

Pakistan, fail to understand infrastructures as historically and socially constituted. Therefore "in naturalizing stagnant water as breeding site—and not infrastructural failure—these intersections foreclose the possibilities for an integrated urban agenda that might address infrastructural disparities alongside disease control" (1470). With this in mind, disparities might equally offer new insights into the unresolved questions that surround the choices made by *Aedes* when settling in Heidelberg.

It is this infrastructural complexity of entangled, transecological networks, socio-material conditions, and shifting spatial complexities that the description of *Aedes* as a "lazy invader" and their fragmented topology might point us to. However, as we come to terms with the diversity of spatial networks and entanglements that are at stake in the distribution of *Aedes* mosquitoes, we still rely on maps and cartographic knowledge that is ill-suited to fully interrogate these spatialities. This leads us to the question of representation itself and the challenge of mapping a combination of territorial, temporal, and bionomic factors. How are we to zoom in on global health challenges related to mosquito mobility without loosing sight of the transecological dimensions of disease control?

ZOOMING IN (WITH *AEDES*)

Maps and cartographic knowledge promise to lay bare a picture of reality that showcases facts about the world. In this sense they have been crucial in exploration projects (e.g., from imperialist voyages to scientific discovery). A new generation of critical cartographers challenge this view through linkages of geographic practice with questions of power and knowledge: "Cartography is being undisciplined; that is, freed from the confines of the academic and opened up to the people" (Crampton and Krygier 2005, 12). Central to this is the participation of citizen scientists and an interested public. Their involvement opens up representations of space and time in cartographies that allow for different rationalities.

It has long been argued that "a cartography of reality must be humane, humanist, phenomenological. . . . It must reject as inhumanly narrow both the data base and subject matter of contemporary academic cartography" (Wood 1978, 207). Such a step is necessary if we are to acknowledge that the production of space goes hand in hand with the production of political subjectivities, which situates maps as social documents that need to be understood in their historical contexts (Crampton and Krygier 2005, 16). Critical cartography expands this perspective with new possibilities for open-source mapping that can also include the more-than-human and in doing so become transecological in character, that is, sensitive to divergent perceptions of space and time.

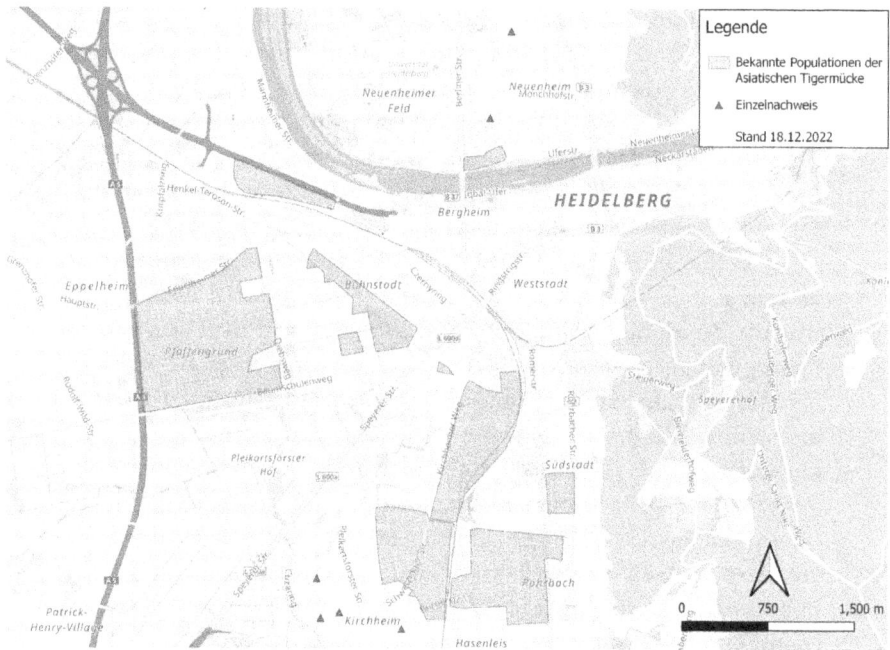

FIG. 9.2. Known populations of *Aedes albopictus* in Heidelberg (individual sightings are marked with a red triangle), March 27, 2023. Source: ICYBAC GmbH—Biologische Bekämpfung von Stechmücken, Speyer (Germany), used with permission.

For the remainder of this chapter, what we aim for by introducing the method of infrastructural go-along is to turn critical cartography into an exercise that decenters anthropogenic perceptions of space and time by embracing more-than-human usage and questions of multi-species conviviality. Given the aim and scope of this contribution we will return to our project and the two divergent ways of cartography we presented above. We have argued that a critical cartography that transcends boundaries between human and more-than-human mobilities can overcome the traditional "cartographic gaze" that coded subjects and produced identities (Pickles 2003, 12). We might think of the current engagement with global sea level rise that has intensified mapmaking practices to rewrite ocean governance in light of the allocation of resources and rights to certain areas. The infrastructure of such marine spatial planning presents a mix of ontological politics, critical cartography, and a critical conceptualization of care (Boucquey et al. 2019). Figure 9.2 might be read as a transitional map that somewhat caters to both, the invasion narrative of mosquito dispersal and also a potential "counter-map" to that of the ECDC that acknowledges, for

Uli Beisel and Carsten Wergin

the 140,000-people cityscape of Heidelberg, a scattered settlement of *A. albopictus* (indicated by red triangles), with no immediate relationship to other territories populated by them.

It is cartographies like these, in figure 9.2, that zoom in and in doing so allow for the representation of a variety of rationalities. What we explore together with colleagues from the Global South in the aforementioned joint research project on human and mosquito mobility is how to further expand such visualizations through the introduction of citizen science and additional aesthetic means. We want to acknowledge the patchiness of *Aedes* distribution that is hinted at in figure 9.2 and in the previous vignette but downplayed by the cartographic powers of the ECDC that present the image of an invasive character (see fig. 9.1). This is what the infrastructural go-along is aiming at, as a method of transecological, multimodal storytelling, in relation to most recent trends in critical cartography.

By combining qualitative and quantitative data, graphical storytelling offers a tool in support of knowledge- and science-based communication that is inclusive of civil society, education, the preservation of cultural heritage, and policy-making (Saxinger, Reinoso, and Wentzel 2022). Participatory mapping has long emerged as a dominating paradigm in approaches to global planning, environmental management, and local community development (see Sletto 2009). While "immersive cartography" is another novel research strategy that integrates environmental art, philosophy, and social sciences (see Rousell 2021), "ecopolitical mapping" has equally been introduced as a comparable technique through which to chart encounters between people and the environment (see Benöhr et al. 2022). The latter was developed by a multidisciplinary group of researchers and artists in Chile. It entails collective cartographic meetings that spell out relationships of power across diverse domains. The infrastructural go-along draws on these collaborative strategies and adds to them a transecological component, that is, the acknowledgment of temporality, processing, and fragmentation in the production of space that is governed by both human and more-than-human agents. The resulting cartographic networks that emerge literally "on the go" open up the investigation to multispecies ecologies of participation. They mark a transition from representative to processual cartographies that highlight the immediate productive uses of territories and infrastructures as site specific and open to divergence, much like in the above vignette.

Alongside geographers, community activists, artists, and new media practitioners, the method of infrastructural go-along lends itself to the exploration of possibilities for modifying cartography. Its multispecies dimension enables eco-political maps to include very diverse processes of usage by opening up map creation to inclusion of the more-than-human. Infrastructural go-alongs thus hold the potential to document the transecological dimensions of mosquito dispersal in relation to human mobility and infrastructure. Drawing on colleagues' work in Chile (Benöhr et al. 2022), participants in related mapping workshops can span from members of an engaged public to social science students, entomologists, and human geographers. They are to map their experiences when sharing space and time with *Aedes* as those significant others that are not able to interpret a map but that are at the heart of what is to be interpreted through mapping, that is, infrastructures, borders and crossings, landscape features, and introduction pathways of infectious diseases. What infrastructures we map will depend on these transecological experiences that time-code the cartography that is being produced. Thus every map that results from an infrastructural go-along is unique, creating a palimpsest of meanings, sounds, images, texts, and other phenomena through which the acknowledgment of entangled mobilities can be made pertinent.

As our method broadens the types of maps that best represent multispecies conviviality, previously overlooked phenomena can be illuminated and a new kind of knowledge built that is based on collaborative research and methods that emphasize the situatedness, processing, and fragmentation of space. This is achieved through greater consideration for underlying aesthetics and sensual experiences, for example, to emphasize processes of building infrastructure and spatial ethnography as critical visualization (Kim 2015). Mapping workshops align scientific inquiry with citizen science. They also enhance multisensory perception and connect issues of place, identity, and affiliation with the development of new cartographic means through which to represent multispecies entanglements. "Community mapping," with its roots in visual anthropology and geospatial and digital media practice, offers another foundation to achieve this goal (Grasseni 2012).

Despite our emphasis on fragmentation, infrastructural go-alongs are part of "mobile ethnographies." They allow for cartographies of larger and heterogeneous sites such as urban areas (see Streule 2020). Furthermore, they have the potential to add to post- and decolonial perspectives on mapping, both conceptually through engagement with sites in the Global South and empirically through encouragement of collaborative work and their emphasis on multispecies networks. Since the resulting maps draw on a diverse set of qualitative and quantitative methods, they equally set a counterpoint to quantified maps and visualizations. We do not want to make an argument here of simplified quantitative data versus rich yet uncategorizable qualitative data (see also Engelmann 2019).

However, we do suggest that a more diverse practice of mapping—pertaining both to *who* maps and *how* we map—might be better suited to the fragmented livelihoods of *Aedes* mosquitoes, humans, and their ecologies. Finally, given the multilayered nature of an infrastructural go-along, let alone each map that is based on a palimpsest of them, the very method must be understood as a product of and contribution to the wider field of digital social sciences and humanities.

Given the particular qualities of mosquitoes, this might equally include mapping sites and environments identified not only visually but through sound. As social media enhance participatory cultures of media production, Droumeva (2017) has shown how communities connect with (environmental) activism in intersecting legacies of acoustic ecology. Incorporating sonic interventions stretches existing cartographies both spatially and temporally, allowing for multispecies perspectives to play a more prominent role. This is particularily relevant for mosquitoes who in everyday experience first and foremost take shape as a mixture of sound (and) bites. Sound in particular is what enables "vernacular" mapping, "a co-fabrication of cartographies by human and nonhuman assemblages" (Droumeva 2017, 338). People are further encouraged to interact with the environment, as they reflect on place in conjunction with its sound qualities. As these "imagined" sonic landscapes take center stage, they push beyond cartographic borders (see Droumeva 2017, 345).

The infrastructural go-along is thus best described as a tool for interconnected knowledge creation that actively works to facilitate the participation of underrepresented groups and challenges prevailing ontologies of space through nuanced, place-based engagement. It seeks to assist the critical engagement with the very idea of mapping as an objective tool of epidemiological knowledge production and a means to control disease risks on a broad scale. The resulting localized critical cartographies of multispecies mobility challenge conventional abstract scale-making practices that undermine the importance of the "situated knowledge," namely site-specific insights related directly to lived experience, environments, and cultures (Haraway 1988).

In the attempts to track, monitor, and control the global distribution of *Aedes*, the infrastructural go-along also engages communities politically and economically, while its design emphasizes the entangled mobilities of humans and mosquitoes and disrupts established preconceptions within global and cross-border health systems. The approach is thus well suited to acknowledge the tremendous resilience that *A. albopictus* has shown in adapting to different geographical and climatic conditions. The mosquitoes take advantage of human-made environments, habitats, and infrastructures. Their ecological needs are well adapted to what humanity has to offer. In response, anthropogenic environments warrant surveillance at points of entry to understand the introduction pathways, causes, and routes of mosquitoes, and to connect these to different aspects of biological and ecological life. While severe biodiversity loss and environmental degradation have stimulated much concern and an engagement beyond academia to successfully confront what generates significant anxieties, the global spread and resilience of *A. albopictus* is told as a very different story. Predominantly, narrations and visualizations embrace a rhetoric of invasion and large-scale dispersal with mosquito populations covering whole regions, nations, and even continents. We have thus proposed the infrastructural go-along as a necessary means to zoom back in on mosquitoes, as the latter turn out to be neither lazy nor invasive, but instead well situated for us to learn more about their expanding transecological habitat.

CRITICAL CARTOGRAPHY THROUGH INFRASTRUCTURAL GO-ALONGS

In this chapter we have been concerned with a search for methodologies that allow for more nuanced, proactive dialogue and critical engagement with mosquito dispersal based on citizen science and the recognition of localized ecologies. We have proposed the method of infrastructural go-along as a means to make multispecies entanglement visible, with an emphasis on the transecological dimensions of the networks they inhabit, catering to both spatial and temporal specificities. For example, the method recognizes the laziness of *Aedes* as a viable phenomenon that deservers further attention through critical cartography and mapping workshops in which researchers, students, and the general public help design representations of mosquito dispersal drawing on diverse aesthetic means that might be both visual and acoustic. In doing so, the infrastructural go-along opens up scientific inquiry into patterns of thought that have been entrenched by an overemphasis on conventional mapping techniques and institutionalized policy and control practices.

Conceived as a mobile ethnography, this method helps us to understand how infrastructures are not mere givens but are shaped through usage. Furthermore, their availability can have transecological consequences well beyond their original intent. They might facilitate and significantly speed up the travel of mosquitoes, as even temporary sites set up for infrastructure maintenance and repair can provide mosquito breeding grounds. The infrastructural go-along puts emphasis on such localized experiences of transecological entanglement to open up new perspectives on site-specific practices that might hinder but also assist the emergence of new health challenges. Related engagement with social and organizational factors that allow for vector-borne diseases to settle in new places might also foster community awareness and participation in mosquito control. In addition, comparative cross-country analysis foregrounds how sociocultural variables—that is, human population density, movement, gender, trade, and commerce—come to matter in different transecological settings.

Faced with unprecedented challenges in the domains of climate change and global health, humanity as an idea is itself in need of rethinking. We need new means through which to map the entangled mobilities of human and more-than-human actors that cater to related risks. The infrastructural go-along enables a transecological, multispecies perspective that seeks to decentralize meta-narratives of widespread invasion by putting human practices and infrastructures in conversation with more-than-human biographies of *Aedes* mosquitoes and their related diseases.

Central to our argument has been the observation that diverse mapping techniques produce competing visions of health security. Ultimately, the representations of disease ecologies remain a snapshot of a wider transecological setting for which control strategies cannot but offer momentary solutions. Since the fast-paced spread of *A. albopictus* is intertwined with international trade and human mobility, we believe that interventions to slow or halt its spread are possible. Its control, however, needs in-depth analyses of multispecies coexistence that is not motivated by a continued overemphasis on security measures that focus on rhetorics of invasion and the eradication of unwanted companion species, but by the transecological coming-into-being of new lifeworlds induced by climate change.

CHAPTER 10

THE UNDERLYING CONDITION AS A DISEASE ECOLOGY

COVID-19 AND THE ANTHROPOGENIC BODY OF INFECTION

Hannah Landecker

We may lack the nerve or imagination to theorise [AIDS] adequately, but it has certainly been theorising us for quite a while.

—*Jean Comaroff, "Beyond Bare Life: AIDS, (Bio)Politics, and the Neoliberal Order," (2010, 22)*

What does it mean for humans to be theorized by a virus that they cannot fully comprehend? In disciplines such as anthropology and the social history of medicine, such a formulation of the "indexical power" of infectious disease is understood as a methodological point: the configuration of illness in society alongside societal responses to epidemics lay bare "economic inequalities, patterns of behavior, and otherwise latent cultural assumptions and discriminations" (Anderson 2010, 252). Diseases, particularly novel ones, therefore work as a revelatory analytic tool or sampling technique for the social scientist. Further, assumptions about causes and the narratives given to pathogens or immune entities or illness processes by the public, patients, physicians, and experimental scientists are fertile ground for analyzing a wide range of phenomena, from health and illness as cultural formations, to the social structure of knowledge communities and biological life as an object of politics and a target of governance.

In this chapter I build on these insights by asking whether the indexical power of COVID-19 may have further dimensions for the social scientist, in that it functions to illuminate anthropogenic aspects of biology that would otherwise be illegible. The analyst obviously must access these properties through bioscientific narratives and descriptions of experiments and findings, and the following analysis is indeed an account of those narratives. Yet this scientific work is situated in "the mangle of practice" in which knowledge making relates in specific ways to the "recalcitrance" of the material world (Pickering 2010; Creager 2002). The material world in this instance is highly recalcitrant to fathoming and control via prior conceptions of infectious disease, and thus there is a lot to learn from following researchers as highly confronting phenomena confound expectation (Dumit 2021). I focus on how that confrontation is raising new questions about the impact of modernization and industrialization on the biology of infection and immunity. These features arise not purely from the minds of observers racing to find a remedy or understand a rapidly changing situation, but from the material weight of the world: data from counties shrouded in unprecedented wildfire smoke mid-pandemic, analysis of lungs taken from deceased smokers, from haywire measurements of blood glucose, and in general from millions of serum samples and blood pressure readings and fluids sampled in the days around inflection points between survival and death in the severely ill.

What is the place of anthropogenic biology in the intersection of chronic and infectious disease so starkly illuminated by the COVID-19 pandemic? The highly variable effects of the SARS-CoV-2 virus on people has indexed a clear vulnerability in bodies and populations suffering from metabolic disorders such as diabetes and hypertension, maladies that in their own right show incidence patterns that track with anthropogenic pollutant exposure, climate change, and industrial food processing, as well as socioeconomic disparity and the social organization of work. In the pandemic, these so-called *underlying* metabolic conditions were clearly tied to much increased risk of severe disease outcomes and mortality in individuals and populations, and these correlations directed intense scrutiny toward the crossing points of chronic inflammation and acute inflammatory crisis, as well as blood glucose control and immune function (Drucker 2021).

Yet this was more than a correlation between acute disease and chronic illness pointing once again to social determinants of health that were perfectly evident already. The search for mechanisms to explain how one human "host" can go without symptoms while another dies of an infection with SARS-CoV-2 has focused unprecedented attention on the conditions that shape the *ground* of infection, on what scientific parlance labels the "in-host ecology": "an ecological environment in which pathogens, host resources and immunity form networks of interacting populations that are analogous to food-webs" (Murall, McCann, and Bauch 2012, 1). In other words, just as predators need prey to eat, and prey need plants and mineral nutrients whose availability are shaped by climate and geography, binding the organisms together in an interacting "food web," human cells, viral particles, and commensal microbes are intermingling and competing in the micro-scapes of membranes, tissues, and fluids in relation to pH, oxygen, iron, sugars, free radicals, lipids, and each other. As with food webs in other "anthropogenic biomes," human consumption patterns and emissions reshape the biotic and abiotic resources and relationships that constitute the body-as-ecology, and thus the possibilities for viral propagation and suppression in that space (Ellis, Beusen, and Goldewijk 2020). Thus the argument unfolded below: in response to events augured in the cells and tissues of patients, a historically distinctive conceptual formation has emerged since the pandemic began, of the body of COVID as an anthropogenic ecosystem.

The idea that the body is an ecosystem is not new, and a shift in the last decade across natural and human sciences characterizing the current era as the Anthropocene has directed attention to the impacts of human activities on all ecosystems via the derangement of the planetary systems that maintain livable conditions (Yusoff 2016). However, the abrupt channeling of both biomedical and public attention through SARS-CoV-2, as well as what this virus indexes as a vulnerable host—in a time of intense focus on the meaning and health functions of the human microbiome—has given specific shape and urgency to these concepts, which are characterized below.

Given the lineaments of this volume of collected papers, I focus in on the themes of ecology and securitization at the intersection of infectious and chronic disease by highlighting the ways in which work on SARS-CoV-2 intensified the question of (1) the human body as an anthropogenic biome in which human industry, emissions, and technologies fundamentally shape both the biotic and abiotic resources with which human cells, viruses, fungi, and bacteria interact, and (2) anthropogenic factors as threats to the security of this bodily ecological territory, in particular its epithelial surfaces, primarily the lung and gut, barriers between the tissues and the outside world. The term *health security* is usually used in relation to states and sovereign borders, and concerns surveillance and control measures aimed at the bodies within a territory or those trying to cross from one to another, yet the assumptions of prior health security regimes have been profoundly challenged by COVID-19

(Lakoff 2021). To bring these concepts into the body is not a metaphorical move. One way that SARS-CoV-2 is "theorizing us" is to indicate the lineaments of unthought forms of health insecurity patterned by the effluents and substances of industrial life mostly ignored by state regulation as all eyes go to the pathogen.

I trace out the viral induction of a theory of the anthropogenic body as an ecology of (highly uncertain) disease control. Obviously, this is a subset of an enormous research enterprise with an equally enormous literature, and what I have chosen to translate into a social scientific idiom out of its current highly technical form in bioscientific discourse is a curated selection focused on the property of boundedness in the anthropogenic body-as-ecosystem. First, I sketch out the epidemiological evidence linking increased mortality and morbidity from infection with underlying conditions. Second, I discuss the mucosal boundaries of the lung and the gut as they are studied and understood in terms of dynamic interactions between microbes and human cells. Third, I turn to the question of how failure in the integrity of the dynamic mucosal boundaries between inner and outer environments is seen and measured not just as a breach of a barrier, but as a degradation of the ongoing capacity to maintain boundedness. Remaining bounded is clearly an issue when pathogens seek entry, and a classic feature of how the immune system has been conceptualized.

Finally, I look beyond the simpler story of enemies versus self. These are boundaries that are always and continuously the product of ecological community structure in the first place, and they regenerate only within the ongoing relationship between many different biological actors that exist on either side of them: the boundary *is* the relation. I show how, in pursuing answers to the questions of why the body of chronic disease is more susceptible to severe outcomes, researchers are detailing how anthropogenic drivers of biology, such as pollution, processed foods, smoke, and endocrine disruptors, make a difference to this "in-host ecology" in which cells and viruses are interacting in relation to bodily boundaries. Anthropogenic factors from emulsifiers to shift work are a part of those relations and thus part of boundary maintenance over time. Failures of maintenance are a driver of "leaky" surfaces and internal chronic inflammatory ground states that lack the capacity either to keep out infection or recover equilibrium in the wake of it.

THE EPIDEMIOLOGY OF COVID-19 AND THE UNDERLYING CONDITION

Even from the earliest days of the COVID-19 pandemic, data emerging from studies of hospitalized patients in China indicated that individuals in intensive care with severe illness from SARS-CoV-2 infection were two to three times more likely to have diabetes, cardiovascular disease, and hypertension than those with milder illness (Li et al. 2020). This led to the early formulation of the idea of the *underlying condition* as a particular risk factor alongside the obvious increased risk that came with advanced age, which prompted further interrogation by the epidemiological community. With some variation, these results have been repeated in populations around the world. Multiple studies of patients in American populations, where rates of obesity and metabolic disorders are very high (according to the CDC, three in four adults are overweight or obese, and half have prediabetes or diabetes), have shown strong associations between body mass index (BMI) and risk of death, and between increasing BMI and risk of other measures that indicate severe illness outcomes, such as intubation or dialysis initiation. As in other studies, these heightened risks with obesity were seen particularly in younger populations under fifty (Tartof et al. 2020; Hendren et al. 2021).

Similar studies in Mexico between February and April 2020 found strong association between obesity, type 2 diabetes, and hypertension and the odds of developing severe COVID-19 (Denova-Gutiérrez et al. 2020). In England a third of all COVID-19-related mortality (23,698 deaths) from March 1 to May 11, 2020, occurred in people with diabetes, meaning that the odds ratio (the odds of death occurring in those with diabetes versus those without diabetes) was 3.51 for type 1 diabetes, and 2.03 for type 2 diabetes (Barron et al. 2020). Worried that studies only of hospitalized patients can unduly skew understanding of these risk factors, researchers using data from UK biobank participants looked at metabolic indicators measured up to eleven years before the pandemic in relation to COVID-19 outcomes, and found that increased serum glucose levels, blood pressure, and waist circumference were specific factors predicting increased mortality (Morys and Dagher 2021).

It is a challenge to narrate this tangle of findings without it becoming a numbing parade of statistics, as each study was done on a different kind of population—hospitalized, diagnosed, diabetic, hypertensive, patients of a health management organization, presence in a health registry or biobank, Brazilian, Italian, American—and often measuring different outcomes, such as some variation of death, ICU admission, or need for mechanical ventilation, which are reported in different statistical terminology, such as odds ratio or hazard ratio. It is difficult to distinguish the vulnerabilities flowing from the conditions themselves, from the potential vulnerability arising out of the pharmaceutical treatments of those conditions, which is often multilayered and excessive in patients with multimorbidity (Ecks 2020). Moreover, studies are occurring at different time points in a fast-moving pandemic of changing viral variants and shifting therapeutic strategies.

Nonetheless, the unfolding parameters of the decisively differential impact of SARS-CoV-2 on people directed investigators' attention toward underlying conditions, many of them chronic metabolic disorders, as an important risk factor. This immediately opened another set of questions about precisely what aspects of these disease conditions were most important in determining disease course. Understanding whether blood glucose control or lipid metabolism or elevated baseline inflammation was the key point of vulnerability was important because each would indicate a potentially different avenue for clinical intervention. Despite argument over the interpretations of what the epidemiology is pointing to, there is a basic consensus, supported by meta-analysis of different studies, that metabolic disorders, particularly hypertension, diabetes, obesity, nonalcoholic fatty liver disease (NAFLD), cardiovascular disease, and kidney failure put individuals at particularly higher risk of severe illness or death from SARS-CoV-2 compared to those living without these conditions, even after vaccination (Stefan 2022).

At the same time the clear socioeconomic shaping of these physiological parameters via work conditions, food insecurity, and access to care highlighted the social and political nature of differential vulnerability at the acute–chronic conjuncture (Manderson and Wahlberg 2020). One striking statement of the meaning of these statistics was an effort to use this welter of epidemiological findings to model the number of COVID-19 hospitalizations in the United States attributable to just four conditions: diabetes, obesity, hypertension, and heart failure, which the authors put at 63.5 percent on average (O'Hearn et al. 2021). In other words, in the absence of these four conditions, the model indicates that far fewer people would have needed hospitalization, even if they were all still infected with SARS-CoV-2. For certain groups and certain diseases, these figures are even more stark; they estimate that among Hispanics age fifty-four to sixty, the largest share of *attributable hospitalizations* arose from diabetes, while for Black adults the largest share of the 70-plus percent was attributable to cardiovascular disease. This is easier, in lay language, to understand. If these conditions were not so prevalent—and prevalence were not so clearly stratified by race and class—the individual and community suffering of so many people in the hospital all at once would not at all resemble what has in fact unfolded.

Thus we see the repeated formulation that the toll that COVID-19 has taken is, at least in part, occurring at the intersection of infectious and chronic disease: the collision of two pandemics, the syndemic of acute illness in a population suffering chronic illness (Stefan, Birkenfeld, and Schulze 2021; Lockhart and O'Rahilly 2020; Singer et al. 2017). Even though this intersection was clearly visible as recently as the H1N1 flu pandemic as well as with SARS in 2003, the scale at which the novel coronavirus cut a swath of disparate outcomes through metabolically disordered populations nonetheless came as a shock to clinicians, scientists, and the public (Louie et al. 2011; Luzi and Radaelli 2020; Singer 2010). In the meeting of the acute and the chronic, the underlying condition is foregrounded and made differently urgent because of its role as the enabling ecology for acute illness. Rather suddenly chronic illness became critical in the short term rather than the long one, with the effect of putting "slow death" on fast forward (Berlant 2007).

BOUNDARIES AS ECOLOGICAL DYNAMICS

Before engaging the question of how the body of infection can be thought of as an *anthropogenic* biome, I must start with exploring how the lung surface—along with other epithelial boundaries, particularly the intestinal lining —is understood as being constituted by ecological dynamics. Much has been made of the specificity of the SARS-CoV-2 virus for the angiotensin-converting enzyme-2 (ACE2) receptor found on the surface of human cells in tissues such as the lungs, pancreas, kidneys, and gastrointestinal tract, yet these cells and their receptors are not just lying openly exposed to the outside air. As with all human compartments, particularly those that interact with the exterior of the body, the upper airways of the lung are protected by a ciliated epithelial lining dotted with secretory cells that produce mucus as well as a liquid film called the airway surface liquid (ASL), which itself is full of antimicrobial molecules and means for sensing pathogens. Such protective fluid and mucus barriers are found throughout the body such as in the gut or the vagina; they are important physical and chemical components of the innate or "constitutive" immune mechanisms that demarcate the inside–outside interfaces in body cavities (Paludan et al. 2021).

After two decades of concerted research on the human microbiome, all such barriers can be thought of as relations as much as they are things. The single layer of human epithelial cells lining these organs has the task of letting some things through but not others, both from outside to inside, and from inside to outside. Specialized protein complexes hold the cells together into a continuous surface, but these cell junctions are far more than a structural glue; they facilitate molecular transportation across the layer and respond continuously to cues coming from the cells in the tissues underneath them as well as from commensal microbes living in the mucus and fluids that line their outside surface. This surface is not sterile, as was long assumed; mucins (sugar-coated proteins excreted by the epithelial cells) provide a stable nutrient environment for a "respiratory microbiome" (Man, de Steenhuijsen Piters, and Bogaert 2017). Thick enough to trap particles and pathogens but fluid enough to be swept back out of the organ via the "mucociliary escalator"—and of course open enough to allow the in-and-out of oxygen and carbon dioxide—this barrier is also full of immune activity. Epithelial cells benefit from the fermented metabolites mutualistic bacteria excrete, such as short-chain fatty acids that cue the production of mucous and the tightening of intercellular junctions.

In short, the bacteria cue the epithelial cells to keep them close by but apart and vice versa. Such dialogic relations are not at all unique to human organs; for example, some species of corn that grow in nitrogen-poor soil have elevated roots dripping with mucilage that hosts nitrogen-fixing bacteria in a similar mutually useful arrangement; slugs and amphibians likewise live in and with their microbiota through mucosal production from their skin (Deynze et al. 2018). More generally speaking, the property of keeping things organized into compartments such that they can interact but not necessarily completely collapse into one big undifferentiated mass is a key feature of life-forms. Indeed, mucosal barriers are the opposite of a complete intermingling. Commensal microbes are very useful when they are kept extremely nearby both in terms of proximity and communication, but not *in* the tissue themselves. Mucus is therefore a paradoxically apart–together boundary substance whose production is cued by the microbes it supports, cues that also tighten the junctions, maintaining the boundary structure. In other words, the boundary does not encircle the ecosystem and demarcate its edge; the boundary is produced in and by the ecological dynamics of mutual feeding and signaling at the surface itself.

ANTHROPOGENIC EDGES

It may not be immediately clear what the dystopic figures in the first section have to do with the apparently utopic boundaries by which various organisms can live closely separated in relations of mutual benefit as described in the second. The question posed by the epidemiological patterns was: Why would *metabolic* dysfunction apparently centered in other organs such as the pancreas and liver be so predisposing to more severe outcomes from a *respiratory* virus? While this has been answered in a number of ways, one explanatory frame has focused on the integrity of bodily boundaries as the mediating factor binding the visceral and the respiratory together, with the boundary microbiome as the "'transducer' of adverse external ecosystems" to internal ones (Prescott et al. 2022, 3503).[1] This focus was not a consequence of COVID itself. Rather, it occurs as a historical conjuncture: a viral pandemic unfolding at a time when a great deal of effort was already underway to understand asthma, irritable bowel disease, and various chronic metabolic disorders in terms of the human microbiota—and its changes under chronic exposure to air pollution and other features of modern life (Kenner 2018).

Obesity, NAFLD, diabetes, and poor blood glucose control are all accompanied by, or perhaps in part caused by, "leakiness" of the epithelial-mucosal barriers between internal tissues and body cavities such as the lung or the gut lumen. Subsequently, leaky barriers lead to a compromised capacity to cope with SARS-CoV-2 infection in several ways. Obviously, viruses can cross damaged barriers more easily. More challenging is seeing chronic low-level inflammatory conditions consequent on boundary degradation as a different kind of systemic insecurity writ large, not just an us-versus-them question of keeping invaders out. Turning to the question of inflammation underscores the *chronicity* of exposures thought to impact boundary ecosystems, from heavy metals such as arsenic or cadmium to endocrine-disrupting chemicals to air pollution. Acute challenges that can be resolved are different from endlessly ongoing low-level challenges, and the mechanisms that support acute responses can be exhausted over time in conditions of chronic challenge (Medzhitov 2021).

Moreover, where and when and how chronically anthropogenic emissions are experienced is patterned by social, economic, and historical factors. Thus we are not speaking of one hypothetical exposed individual versus another, less exposed one, but of lung and intestinal surfaces and bodily inflammatory states of populations showing gradients across large numbers of people due to social conditions such as the proximity of neighborhoods to freeways or to past and present sites of industrial activity. Viruses by their nature do not simply stay in the bodies they infect, and the ecological dynamics of viral propagation of any one host will of course impact everyone around them to varying degrees. In other words, anthropogenic edges do not stay at the edges; the underlying condition is a disease ecology both inside the body and between bodies.

These rather abstract points should become clearer in working through a few examples. Let us begin with the simpler case, in which different anthropogenic factors have been identified as changing the security of the mucosal, cellular, and immune barrier against SARS-CoV-2 entry into the body in the first place, before moving on to more complex ideas about the long-term shaping of inflammatory ecologies. As discussed in the introduction to this volume, air pollution has come under particular scrutiny as a risk factor, in no small part because of historically unprecedented wildfire seasons co-occurring in North America and Australia with waves of viral infection as part of the COVID-19 pandemic (Curtis 2021). Initial epidemiological correlations between air pollution indexes and confirmed COVID-19 cases, or between particular particle pollutant densities and mortality, were seen in northern Italy and China (Fattorini and Regoli 2020; Zhu et al. 2020). These initial studies were accompanied by the hypothesis that small particle pollution was contributing to the mechanical delivery of SARS-CoV-2 virus into people's lungs. At its simplest, this is the idea that industrial toxicants carry injury into the biological body: modern life meets the natural state of the human lung, and injures it.

Yet this basic model was rapidly made more complex when the cellular membrane protein ACE2 receptor was identified as the specific portal of entry that SARS-CoV-2 uses to latch onto and get inside human cells (Hoffmann et al. 2020). Viruses that get inside cells are the ones that will be replicated and go on to infect more cells. This puts considerable evolutionary pressure on the specificity that a virus has for a common cellular structure. As ACE2 is essential for organismal life, carrying out important functions to do with the regulation of blood pressure, human cells cannot do without it, and indeed, cells use ACE2 to protect themselves against lung injury. Chronic exposure to air pollution upregulates (increases) the number of ACE2 receptors produced by lung epithelial cells (Samavati and Uhal 2020). This presents us with the conjuncture between infection and pollution: a virus that has evolved specificity to a molecule that will be present in increased quantities in lungs exposed to chronic air pollution or cigarette or wildfire smoke. People who breathe in more pollution for longer could be more likely to have higher numbers of potential viral entry points on the cells lining their lung surface. And then, of course, the more there are of these people who do not manage to evict the virus before it enters their cells, the more viruses there are being breathed out. This is one possible explanation for the correlations between air pollution and infection rates seen in the environmental health literature, and thus a component of the "atmosocial" milieu of COVID-19 discussed in this volume (see Opitz, chap. 11).

The lineaments of an anthropogenic biology *in which an organism's adaptation to higher levels of pollution becomes the biological condition shaping viral evolution* thus begins to come into view. This is not a failed boundary but a change in the condition of the surface. Indeed, a cluster of explanations or hypotheses have focused on the "modern" lung surface, that is, the one conditioned by contemporary lifeways over time, not just the one focused on ACE2. For example, epidemiological and clinical indicators of abnormally high levels of blood glucose (hyperglycemia) as a risk factor for severe disease outcome has trained attention on glucose at the lung surface. In a remarkable analysis based on machine-learning processing of the COVID-19 Open Research Dataset of over 240,000 research papers on COVID-19, Emmanuelle Logette and colleagues write that glucose was the entity that emerged in their mapping of the clinical and experimental literature as showing "the deepest and broadest association with all stages of virus infection," and that even at the level of this first interaction between lung and virus, elevated glucose could matter in multiple ways (Logette et al. 2021, 6). They start by discussing the very meeting point of virus and lung surface.

Airway surface liquid is a term used to describe the combined structure of a thin mucus layer overlying a periciliary fluid layer closer to the lung epithelial cells. This double mucus-liquid layer has to be thick enough to trap inhaled particles and pathogens, and fluid enough to allow the cilia to beat and propel these unwanted entities out of the lungs (on the "mucociliary escalator"), yet chemically spare enough to not foster bacterial or viral life. The glucose concentration in the ASL is "especially carefully regulated" at levels ten to twelve times lower than the blood (Logette et al. 2021, 18). High glucose levels in the ASL are seen in diabetes, hyperglycemia, and obesity, which themselves are causally linked to highly processed foods, endocrine disruptors, and shift work. In addition, conditions that affect the tight junctions between the lung epithelial cells, such as asthma or exposure to cigarette smoke or pollution, also result in the loss of regulatory control over the diffusion of glucose from the tissues to the surface fluid.[2]

Elevated glucose levels affect the innate immune mechanisms in the ASL.[3] In other words, glucose availability in the ecosystem in which a whole range of interacting entities are living will impact the way they interact. For example, lung epithelial cells secrete mucins into the mucus layer that coats them, as well as antimicrobial proteins, such as C-type lectin. C-type lectins are characterized by a specific carbohydrate-binding domain that can selectively hold on to structures called glycan trees that protrude from the viral lipid envelope. Lectin binding gathers a number of viruses together in a clump and hinders the capacity of the virus to bind to host cells, as well as flagging down resident macrophages and other immune cells to phagocytose (engulf and break down) the viruses. The specificity of the interaction arises from the chemical structure of sugars. C-type lectins bind glycan trees sticking out from the viral envelope. High levels of glucose in this normally sugar-sparse terrain compete for the binding sites on these lectins, diluting their efficacy in sticking to and agglomerating viruses.

THE ANTHROPOCENE AS CHRONIC INFLAMMATION

The examples we've just worked through touch on two ideas about anthropogenic biology at the lung surface, one focused on pollution-impacted, receptor-mediated viral entry at the human cell surface and another focused on the glucose environment that microbes live in at the lung surface. Their relative contribution or additive interaction is unknown, but they are not mutually exclusive. Whether or not these are the most correct hypotheses to meet our current moment is not the question. Rather, notice how their conceptual shape involves a pattern of attention to the manner in which the human-made external environment (polluted air, industrialized diet) is part of the constitution of conditions at the bodily surface and thus the microecology in which tissues, components of the immune system, and viruses interact. These conditions are in turn evolutionary selection pressures that shape what viruses can do and where they successfully replicate or don't.

These conditions of the surface are also the conditions of the insides. In this third and final example, I take up the attention the COVID-19 pandemic has directed toward chronic inflammation as an anthropogenic condition (Landecker 2024). The relationship between virus, inflammation, disorder, and the anthropogenic barrier runs like this: much of the severe illness in COVID patients is caused by the body's own immune inflammatory response to infection, in an overwhelming "cytokine storm" in which signaling molecules that function to ramp up different components of the body's response to infection are overproduced and drive tissue damage and organ failure. The epidemiological connection between COVID and diabetes or fatty liver disease is accounted for in part by chronic inflammation that is already present in the body long term (Vasbinder et al. 2022).

Chronic inflammation in turn is linked to, among other factors, breakdown at epithelial barriers, primarily studied in terms of "leaky gut," the gut wall being a mucosal boundary much like the lung surface. Increased bacteria and bacterial antigens moving from the intestinal tract into the bloodstream drive low-level but ongoing heightened activity of immune cells such as macrophages in the liver, which send out signals to other immune cells, keeping them at a chronically high-alert level. Finally, leaky gut and disruptions of the skin epithelial barrier such as asthma and dermatitis have risen sharply as populations are increasingly living with long-term exposures to a suite of anthropogenic factors, from emulsifiers in food to enzymes in detergents to particulate matter and nitrogen dioxide concentrations in the air (Fiorito et al. 2022).

In this late industrial time we should perhaps not be surprised by this reconceptualization of disease as a failure of maintenance, in this case of the dynamic relation that is the organismal epithelial-mucosal boundary (Fortun 2012). Indeed, the maintenance of organized spatial compartmentalization through robust semipermeable barriers around cells, organelles, and inside–outside surfaces of the body is increasingly articulated as an important "hallmark of health" (López-Otín and Kroemer 2021). Yet it is worth paying attention to the details by which researchers are thinking about this particular failure of maintenance: "how adverse macroscale ecology in the Anthropocene penetrates to the molecular level of personal and microscale ecology, including the microbial systems at the foundations of all ecosystems" (Prescott et al. 2022, 3498).[4] Such an approach sees the Anthropocene—a term denominating the period following the Holocene in which human activity becomes the primary driver of geological and meteorological change—as also taking place at the gut wall, in the ecology of liver macrophages, and through the tightness of protein bonds. It is not just that changing climates or shifting nutrient conditions creates the environment *in which* organisms live, but that the micro-scapes of life are re-territorialized by anthropogenic abiotic and biotic features (D'Abramo and Landecker 2019).

For example, various commercial emulsifiers—substances that keep one liquid suspended within another, such as keeping oil and vinegar mixed in salad dressing—demonstrably change the viscosity of mucus by changing the interactions between the constituent glycoproteins that compose it, thus altering the rate at which bacteria can move through it (Bancil et al. 2021). Such mechanical questions of viscosity are particularly important in the colon mucosa, a dense mucous layer in which microbes are generally not present, lying between the epithelium and the looser mucous layer in which complex communities of commensal bacteria and fungi live. Emulsifiers are just one example of the substances in highly processed foods that have been shown to "potentiate" intestinal and systemic inflammation via disorganization of this carefully orchestrated set of mutualistic relations (Chassaing et al. 2017).

It may seem to rob the Anthropocene of some of its grandiosity to bring it down to banalities such as salad dressing composition or the chemicals released in the browning of potato chips, but it is exactly in such details that we are able to see the shape that questions of bodily security are taking in this discussion of chronic inflammation. While the examples provided in the previous section focus on viral entry and capacity to persist in the lung microecology, here the loss of security is a degradation of boundary maintenance, not under extreme conditions but under very mundane and ongoing ones. The condition of daily living becomes, over time, the loss of the capacity to maintain homeostasis, a loss that is starkly illuminated by the infectious event. While it remains unclear whether factors such as gut permeability causally precede obesity and fatty liver disease or vice versa, it is evident that these conditions reinforce one another and lead to chronically elevated inflammatory responses. Moreover, researchers have proposed a "gut-lung axis," constituted by a flow of intestinal microbiome metabolites through the circulation of blood and lymph, that makes inflammatory conditions in the gut such as irritable bowel syndrome cause respiratory symptoms in the lung (Mateer et al. 2015).

Inflammation is classically described in medicine as the acute reaction to injury, in which tissues swell, hurt, and redden—tumor, rubor, dolor, and calor. The kind of inflammation seen in chronic metabolic disease is by contrast not necessarily immediately sensible on the part of the sufferer. The breakdown of the organization described above, in which the mutualistic relationship of the microbes and human cells is maintained by the structure of a mucosal layer that keeps the two very close together yet spatially distinct, results in an elevated presence of bacteria in bodily tissues, particularly the liver. The liver is where 80 percent of the body's immune macrophages reside, and the sensing of and response to bacterial presence drives the production of the same cytokine signals that in an acute situation rise up and marshal immune cells to a site of injury. Yet this immune response is endlessly being cued without the original signal of injury or infection becoming resolved.[5]

Whether researchers are focused on the boundary ecology of the mucosa or looking to other sources of chronic elevated inflammatory signaling, such as an expanded population of adipose cells in fat tissue, the social causes of these inflammatory states are front and center. For example, proponents of the "epithelial barrier hypothesis" are unabashed about positing that "laundry and dishwasher detergents, household cleaners, surfactants, enzymes and emulsifiers used in the food industry, cigarette smoke, particulate matter, diesel exhaust, ozone, nanoparticles, and microplastics" are the key drivers of the inflammation that has become the ground state in a large percentage of contemporary populations (Akdis 2022, 41). And this is the ecology into which the SARS-CoV-2 virus arrived, both in the sense of the bodily ecology and the knowledge ecology responding to these changes in the material states of bodies under clinical care.

Well before the COVID-19 pandemic, the link between chronically dysregulated blood glucose, excess adiposity, anthropogenic environmental influences and the chronic low-level activation of the innate immune system was being made (Wakefield-Rann 2021). It had been noted in smokers that the chronic activation of macrophages "is paradoxically associated with attenuated innate immune responses to both bacterial and viral infections" (Gleeson, Roche, and Sheedy 2021, 629). While one might think that if the macrophages are chronically activated, they would be all the more ready to meet an additional challenge when viruses come along, actually this activated state as a chronic baseline seems instead to lead to "bioenergetic exhaustion" of the cells, tamping down their ability to respond to acute challenges.[6]

The COVID-19 pandemic has not so much created as highlighted certain elements of a conceptual landscape built out of a world of increasing anthropogenically driven metabolic disorder: "the growing notion that obesity-associated morbidity is related to its temporality (i.e. chronicity), and that 'physiological inflammation' is likely a normal acute adaptive response to short-term changes in energy balance, which become pathogenic and maladaptive when the metabolic imbalance becomes chronic" (Lahav et al. 2021, E711). Instead of the periodicity of rise and fall, we get cytokine storm. What we see here is the particular time structure of the Anthropocene of cellular processes, after homeostasis. The Anthropocene articulated this way is more than a time period that comes after a change in conditions: it is a time structure in which periodicity is lost. In this time after homeostasis, challenges (of nutrient excess or shortage, of stress, of toxins) that used to occur *only sometimes* and were dealt with by acute responses with a quick return to baseline, are *now always*.

VIRAL APPERCEPTION OF THE HUMAN CONDITION

It is not a given that humans are able to know how molecular topographies are contoured by society, culture, history, industry, and politics, just because these changes are consequent on human institutions. Viral apperception (for lack of a better word) of the human condition is on its own terms and in its own ecological and evolutionary space: what viruses bind to or pass through will provide an unanticipated picture of human life at the molecular scale. That it takes a nonhuman agent to illuminate and detail the dimensions of anthropogenic biochemistry to humans may seem paradoxical, but of course one should never mistake effects that issue from human activity for being synonymous with things completely comprehended by human actors. From this perspective, the anthropogenically contoured living surface—if one were to imagine the world's epithelial surfaces as a fairly continuous plane from an airborne viral point of view—is as much a part of the event that is the COVID-19 pandemic as the viral pathogen.

In this essay I have described an *ecological body of infection*, a distinctive shape of thought arising from observation of the cytokines, clotting

Hannah Landecker

factors, and blood glucose levels of millions of afflicted bodies. As we have learned from philosophers of viruses—and see in other chapters in this volume—ecologies make the difference, not the thing itself; yet not enough account has been taken of the sociohistorical nature of those ecologies (Méthot and Alizon 2014). It is my hope that this analysis allows readers to think about anthropogenic microecologies more broadly. What the SARS-CoV-2 virus indexes, among other things, is the extent to which the underlying condition is not underlying at all, but is a historically distinctive anthropogenic body of infection, a microecology of late industrial times, bearing with it the paradoxical sense of life in the wake of modernity as ubiquitous exposures leading to *uneven effects everywhere.* These may be microscopic topographies and temporalities, but they are en masse and are determining the viral course through populations and time. The viral "reading" of metabolic landscapes in COVID-19 occurs at a scale of the simultaneously collective and microscopic, forcing us to think within and across the cells and cellular processes of the planet, to "reason from organisms' metabolisms outwards," rather than starting at the planet and trickling down to the organisms caught up in the currents of how human activity is changing the environment (Lenton, Dutreuil, and Latour 2020, 248).[7] Cell and microbial ecologies are not downstream receptacles of events that happen elsewhere, but are places of cellular weather that we might understand as Anthropocene events, just as unprecedented downpours and receding glaciers are.

CHAPTER 11

THE ATMOSOCIAL CONDITION

ECOLOGIES OF BREATH AFTER COVID-19

Sven Opitz

In an open letter published on July 6, 2020, 239 scientists from thirty-nine countries declared that "it is time to address airborne transmission of COVID-19" (Morawska and Milton 2020). They demanded that scientific evidence showing that "viruses are released during exhalation, talking and coughing in microdroplets small enough to remain aloft in air and pose a risk of exposure at distances beyond 1–2 m beyond an infected individual" be taken seriously (Morawska and Milton 2020, 2311). The letter was primarily directed at public health authorities worldwide, criticizing their reluctance to acknowledge the role aerosols play in spreading disease. Most notably, not even the World Health Organization (WHO) recommended any general precautions against airborne infections at the time (see WHO 2020a). A WHO tweet on March 28 implicitly accused those claiming that airborne contagion was possible of spreading misinformation. Under the heading "FACT CHECK: Covid-19 Is NOT Airborne," the claim appears stamped with "INCORRECT" in the tweet (WHO 2020b). In the initial phase of the pandemic, the possibility of infection via the inhalation of aerosols was sidelined by two other modes of transmission: infection through the exposure of droplets to the eyes, nose, or mouth, or through contact with contaminated surfaces. These modes of transmission have, in fact, dominated public health reasoning and the epidemiological imaginary throughout the entire twentieth century (Jimenez et al. 2022). Airborne transmission, in contrast, has long been regarded as marginal, limited to a comparatively small number of infectious diseases (tuberculosis and measles being the most prominent among them).

However, the status of aerosols has changed since the summer of 2020 (Baraniuk 2021; Greenhalgh et al. 2021a). The controversy has rendered them a public thing that has slowly transformed from a matter of vital concern into a matter of fact (Latour 2004). Public health institutions, from the WHO to national authorities, such as the Centers for Disease Control and Prevention (CDC) in the United States or the Robert Koch Institute (RKI) in Germany, have adapted their guidelines and fact sheets. Mounting epidemiological evidence for airborne transmission—supported by steady news coverage—has drawn public attention to the issue. According to Nortje Marres (2005, 217), "implication in an affair is what sparks public involvement." During the COVID-19 pandemic, virtually every breathing human body was brought into an "affair" with the aerosols around it, changing the understanding of our surrounding air from an implicit background into a highly fraught entity. It could no longer be taken for granted that ambient air could be safely consumed. Rather, the need to breathe in and out became a risk interval: the necessity to breathe out produces a risk for others, the necessity to breathe in puts the body at risk. In this way COVID-19 rendered ecologies of breath deeply problematic. Atmospheres—in the literal sense of breathing spaces (see Garnett 2020; Mitman 2007)—appeared as indispensable for life and as pathogenic at the same time. The "immersion of the living into an atmospheric milieu"[1] (Sloterdijk 2004, 103) has become tantamount to the immersion into a milieu ridden by the possibility of viral encounter.

This chapter argues that the atmospheric space that publicly emerged during the pandemic also contains a lesson for sociology. Paying close attention to the controversy over airborne transmission can contribute to the understanding of air as a social medium and thereby to attune the social to its "elemental" dimension (Engelmann and McCormack 2021). With COVID-19 breathable air could no longer be relegated to an *Umwelt* neatly separated from social interactions. The debates on airborne transmission show that the social not only proceeds *in* atmospheres, but *through* atmospheres. The pandemic therefore offers the opportunity to understand the social as part of the atmospheric composition of breathing spaces: the social proves to be materially implicated in their very makeup and, in turn, is affected by them. This view builds on Tim Ingold's observation that the "very essence of conviviality lies in the sharing of breath" (2020, 162). Collective breathing is a constitutive part of speaking and listening, gesticulating and acting—*on* or *with* each other, or sometimes merely *in proximity to* each other. Collective breathing in shared atmospheres is part of a multitude of social practices such as shopping, using public transportation, standing in elevators, visiting museums, or going to the gym. In all of these settings we experience today a health risk that emerges from our bodily "being-with" (Nancy 2000, 61) and "being alongside" (Latimer 2013). SARS-CoV-2 thus highlights the infectious quality that the social has acquired through the aerosol pathway.

So far, however, the implications that the aerosol pathway, and the risks of collective breathing, bring for the notion of the social have not been properly theorized. To unfold them this chapter introduces the concept of the "atmosocial." With it I seek to render productive the tension that exists between the ecologizing move happening in the governance of health and in social theory identified in the introduction to this volume. On the one hand, daily concerns about breathing indoors and air safety show the extent to which current modes of infectious disease control have to account for elemental surroundings. On the other hand, attending to them may change how we conceive of relatedness itself. Even if both dimensions cannot be neatly separated, this chapter seeks to highlight the latter, aiming to think of the social through ecologies of breath. For this purpose it follows the problematization of aerosols during the COVID-19 pandemic.

The analysis proceeds in three steps, each of which will elaborate on a key feature of the atmosocial. The first section shows how the notion of airborne transmission ruins measures of securing social proximities through what Erving Goffman calls "territories of the self" (1971, 28). The atmosocial takes a voluminous shape; it exceeds surfaces and modes of ordering premised on gravity. The second section proposes the understanding of atmospheres as turbulent media that link fluid dynamics with affective properties. The atmosocial is made up of intensities that are volatile, elusive, and uncertain. The third section focuses on the permeability of bodies in atmospheric milieus. As will be demonstrated, the atmosocial exhibits a form of relationality that differs fundamentally from established models of relatedness, such as the network or the classical concept of interaction, both of which confer notions of bounded entities. The conclusion reflects on the particular form of "theorizing" (Swedberg 2016) conducted throughout the analysis. The chapter opens a "trading zone" (Galison 2010) that brings together disciplinarily heterogeneous forms of knowledge across the divide between social and natural sciences. To outline the contours of the atmosocial, these knowledge forms are read through one another.

TERRITORIAL ORDERING AND ITS LIMITS

In the repertoire of infectious disease control measures applied during COVID-19, the injunction to stay six feet apart remained in place throughout the pandemic in many countries. The spatial norm materialized in the form of floor markings and barrier tapes, but also verbal reprimands when someone's physical distance was deemed too close. Whereas a "coughing etiquette" (Brown et al. 2021)—turning one's head away from other people, cupping the mouth with the hands, coughing into the elbow, among others—had long been part of hygiene behavior, COVID-19 consolidated a supplementary "etiquette of spacing" (Brown et al. 2020, 973) that human bodies were expected to follow in relation to one another. In conjunction with calls to abandon the greeting ritual of shaking hands and instead perform contact-free equivalents, such as bowing or raising a hand, distancing rules were a substantial intervention into the interaction order (Jensen 2021).[2]

Drawing on Erving Goffman's (1971) sociology of "relations in public," we can describe this development in territorial terms. Goffman adopted the notion of territory from ethology to illuminate the intricate organization of social interactions: individuals demarcate "territories of the self" by using markers that express claims about who can approach them and in what fashion, either directly through bodily proximity or touch, noises or looks, or indirectly through positioning things (Goffman 1971, 41). Territory in this sense is a social ordering mechanism that operates by "carving the environment" (Brighenti 2010, 61). It inscribes norms about how to mutually relate into spatial renderings. Against this backdrop, the distinctive feature of the physical distancing rule under COVID-19 is its almost universal validity across very different situations. Goffman's "egocentric preserve," understood as fixed to an individual and thus moving around with them (1971, 29, 255), was supposed to have the same measurement in any public setting, irrespective of the particular person involved. The German government integrated this general rule of conduct into the so-called AHA-formula that offered guidance for every daily interaction. While the *H* in the formula refers to hygiene and the second "A" to wearing masks (*Alltagsmasken*), the first *A* stands for *Abstand* (distance). As the formula illustrates, COVID-19 had given rise to a territorial meta-convention that remediated context-bound norms.

Crucially, the safety promise connected to this territorial ordering was challenged in the controversy about aerosols because it had been based on the premise that SARS-CoV-2 travels via droplets. The distinction between aerosols and droplets often refers to the size of the particles, but it is one that has become a matter of controversy itself. While some public health institutions, including the WHO, set the dividing line at 5 micrometers, scholars from atmospheric science advocate for aerosols to be classified not for what they *are*, but what they *do*. They distinguish them through their performances, not essences. At the top of their "list of actions" (Latour 2014, 3) is the ability of aerosols to enter humans' respiratory tracts. Particles can be inhaled up to a size of 100 micrometers. Equally important, aerosols and droplets perform different styles of movement when suspended in the air. Droplets fall to the ground after a short time span, which also implies a limited horizontal range. Aerosols, in contrast, are "ideally airborne particles," as Martin Kriegel, director of the Hermann Rietschel-Institute at Technische Universität Berlin, puts it.[3] This means that they defy gravity. Like dust or cigarette smoke, they can travel across larger distances and stay afloat in the air for longer time spans.[4] As a precautionary measure, physical distancing is therefore primarily suited to the behavior of droplets. For aerosols, however, though not completely useless, its protective capacity is deficient.

Since the onset of the pandemic, several epidemiological case studies have been published on aerosols in action. They shed light on scenes of infection and, by documenting the significance of airborne transmission, re-problematize spatial relations between human bodies. An outbreak event on January 24, 2020, at a restaurant in Guangzhou, China, is one of the first documented cases where the infection could not be fully explained by droplet transmission, as it was detected in people sitting at several tables with a distance between them of up to 4.6 meters (Li et al. 2021, 7). Video camera footage showed that the infected persons did not have close contact nor spoke with each other. The subsequent simulation of the event with a tracer gas detected the interplay of two factors, both related to the circulation of air: very little supply of fresh air due to low ventilation rates and the air conditioning forming a "contaminated recirculation bubble" (Li et al. 2021, 7) that enveloped the affected three neighboring tables. Together these accounted for high virus concentrations in the room. Ventilation practices (or, rather, the lack thereof) in conjunction with the technological production of a comfortable, well-tempered environment, and a particular way of using indoor space, led to an infectious atmosphere.

The investigation of an outbreak at a meat processing plant in Germany in May 2020, widely reported in the media, even documented a transmission from a distance of over 8 meters (Günther et al. 2020). In addition to stressing the role of low air exchange rates and indoor air-flow, the authors add "climate conditions" (Günther et al. 2020, 1) to the set of factors that contribute to contagious atmospheres. This includes temperature, with environments cooled down to approximately 10°C increasing the viability of the virus. Moreover, and in accordance with reports on outbreaks in singing choirs (Naunheim et al. 2021), the study highlights that heavy breathing, in this case due to physically demanding work, can increase the risk of infection.

Both case studies clearly show that airborne transmission cannot be stopped by attempts at territorialization. Territories need to be demarcated, yet the inscription of boundaries into air is notoriously difficult. Verbal announcements perpetuate and increase the risk of infection since the air is both the medium of communication and of communicability (see Horn 2018). In fact, air has de-territorializing properties that complicate practices of territorial distancing. The challenges at stake are well known from the history of warfare: any deliberate release of biological or chemical agents into the air must reckon with the possibility of changing winds.[5] Indeed, it has proven difficult to confine pathogens to the dividing lines between friends and enemies. In a similar way, it is hard to enclose those aerosols that are continuously released through breathing, talking, or singing. Partitioning devices made of acrylic glass can stop droplets, even if they do not seal off a space completely. In contrast, aerosols tend to transgress boundaries, at least in the absence of socio-technical equipment such as "air purifiers" or UV-C radiation devices. Consequently, over the course of the pandemic we have added practices to our everyday routines that increase the volatility of the medium: we open windows and switch on ventilation. Confronted with such air-flow conditions, the concept of territory exhibits limitations. As Goffman has succinctly remarked, the territory is "a contour, not a sphere" (Goffman 1971, 30). If sociology wants to take up the provocation of the aerosol for conceiving of the atmosocial condition, it has to move from the territorial to the atmospheric.

TURBULENT ATMOSPHERES OF FLUIDS AND AFFECTS

The atmosphere requires a transition into three-dimensional space. It adds a vertical dimension to the horizontal plane on which communicative action and social interaction are usually situated (S. Graham 2016). This verticality is seldom taken into account in the social sciences, and then almost only in diagrams about domination or inequality, sovereign power or stratification. In comparison, the ecologies of breath COVID-19 has forced us to grapple with constitute a thick milieu in which bodily relations merge with vapors and effluvia. Accordingly, the pandemic has considerably spurred societal concerns to secure atmospheric volumes (see Elden 2013). This has proven difficult, however. Not only do human bodies' own emissions always affect the atmosphere around them; ecologies of breath are always in motion and changing their shape, making their circumference hard to determine. As this section will elaborate in more depth, atmospheric volumes are irreducibly infused with uncertainty. Their aleatory composition defies volumetric certainty; hence the turbulence of fluids intersects with the turbulences of affects.

The urge to gain better knowledge about the spread of SARS-CoV-2 has motivated physicists to study the fluid dynamics at work in a contagious interaction. In the words of Lydia Bourouiba from the Department of Civil and Environmental Engineering at MIT, the aim is to "understand the mechanisms by which fluids shape the dispersal and transmission of pathogens" (2021, 474), that is, their "encapsulation" in a host, their "emission," their ecological "persistence in the environment," and the "new-host infiltration" (476). Modeling this process, the focus lies on the "fully turbulent flow" of a "puff cloud of hot and moist air" released by breathing bodies (480).

For social scientists turbulent fluids are good to think with for at least two reasons. First, the dynamics described by Bourouiba occur exactly at the scale of the atmosocial. This sets them apart from the scales at which molecular biology and epidemiology operate—the two knowledge formations that have dominated scientific thought on infection for more than a century. The dynamics of turbulent fluids occur in between the scales of microorganisms and human populations. In a sense physicists like Bourouiba model "breathed communes" (Sloterdijk 2011, 17) as shared spheres that materialize through breath.[6] Furthermore, turbulent fluids offer a useful model of order and disorder. Michel Serres (2018, 21–26) suggests we think of turbulence in terms of the Lucretian notion of *clinamen*: the small deviation of moving matter from a linear path that results in chance collisions or encounters. The unpredictable deviation, or swerve, accounts for turbulent interactions. The moment in which the "parallel flow" of matter transforms into a "swirling flow" (24) is the "inception of turbulence" (25). Serres underlines the unpredictable, spontaneous nature of the clinamen: "the incident is uncertain in time, uncertain in place, and in each case undetermined" (25). This does not simply amount to a disordered state. The fluctuation of the swerve rather adds a stochastic element to the constitution of order and thereby an irreducible degree of uncertainty.

This has important implications for everyday encounters in, and with, atmospheres transmitting airborne diseases. One never knows precisely where a potentially virus-laden cloud ends. In general, clouds have fuzzy boundaries. Yet they not only envelop a speculative quality regarding their exact whereabouts or their very composition; due to their turbulent state, clouds form heterogeneous mixtures that constantly evolve, which makes them hard to grapple with. Of course, during the COVID-19 pandemic several attempts were made at calculating how airborne exposure took place. For example, carbon dioxide measurements were used as a proxy to infer what amount of rebreathed air posed an infection risk. Widely distributed newspapers put models of aerosol dispersion on their websites, allowing users to simulate different socio-spatial settings such as classrooms, restaurants, or offices.[7] However, the multiplicity and unpredictability of influential factors—from airflow to the thermal impact of body temperature or even energy sources like lightbulbs—prevent an exact knowledge of what the dynamics of airborne disease transmission are.

The speculative quality of an encounter in and with pathogenic atmospheres points at the the intimate link between the elemental composition and the affective charge of atmospheres (McCormack 2018, 20–21). Their indissoluble liaison characterizes the atmosocial. Scholars from the humanities have emphasized in recent years that atmospheres are not only extensive but also intensive entities. They can allow bodies to breathe freely or cause them suffering with their oppressive climate—up to the point at which people feel they might suffocate. Such atmospheric effects are not to be understood as projections of an emotional subject. Instead, atmospheres emanate from the relational interplay of the elements involved (see Anderson 2017, 137–62). According to Gernot Böhme (1993, 121), things and bodies of all sorts suffuse a situation and produce "an indeterminate spatially extended quality of feeling" (118). They not only "tune" but "tincture" (121) their environment. This description resonates with the etymology of infection that harbors the meaning of "dyeing" or "to stain something in the sense that it becomes tainted, spoiled or corrupted" (Temkin 2006, 457). By stating that the atmosphere is "something which flows forth spatially, almost something like a breath or a haze," Böhme (1993, 117) invokes epidemiological notions like miasma and effluvia. Until the late nineteenth century "miasma" designated emanations of all sorts of putrefied matter while "effluvia" referred to the pathogenic vapor around sick bodies (see Mukharji 2012, 311–12). The latest controversy around virus-laden aerosols has brought us, again, into closer proximity to such atmospheric models of transmission, despite the fundamental difference in the underlying theories of disease causation.

The atmospheric link between actual breathing spaces and affective envelopment appears prominently in the controversy about the different modes of infection that took place after the rise of bacteriology at the turn of the twentieth century. This controversy is highly relevant for a "history of the present" (Foucault 1997, 30) because it cemented the sidelining of airborne transmission in favor of droplets and contaminated surfaces. At the time several experiments had provided evidence that bacteria could indeed stay in the air for prolonged periods and travel several meters (Polianski 2021). However, looking at historical sources, it seems that the possibility of "volatile contagions" was mainly rejected due to concerns about its interference with established public health measures. In his influential book *The Sources and Modes of Infection*, Charles Chapin (1912), who would later become president of the American Public Health Association and who was characterized in 1967 by Alexander Langmuir as "the greatest American epidemiologist" (cited in Jimenez et al. 2022, 7), discards the possibility of airborne transmission. He does so mainly based on an affective calculus: "Infection by air, if it does take place, as is commonly believed, is so difficult to avoid or guard against, that it discourages effort to avoid other sources of danger" (Chapin 1912, 314). Chapin does not deny the existence of "volatile contagions" via air. Yet the threat of infection would be so ubiquitous and pervasive that people would be unmotivated to take precautions, such as keeping distant. For Chapin, physical distance is the most effective measure against what he considered the dominant transmission path of "contact infection," which includes infection both via droplets ("sprayborne") and contaminated surfaces ("fomites"). "If the sick room," he rhetorically asks, "is filled with floating contagium, of what use is it to make such an effort to guard against contact infection?" (314). He then concludes that "it will be a great relief to most persons to be freed from the specter of infected air" (314). The affective relaxation vis-à-vis the atmospheric visitations goes hand in hand with the aim of generating commitment to measures against "contact infection."

Current debates in social and cultural theory have highlighted the nonrepresentational quality of atmospheres (see Anderson and Ash 2015). Atmospheres are not primarily structures of meaning. They assume a felt presence that is bodily sensed. Sanitary assumptions about relations between smell and health in the modern industrial city are a case in point (Kiechle 2017). The perception of foul airs, stenches, and olfactory nuisances have generated atmospheric anxieties about "safe breathing" (Kiechle 2016, 757). Until today, olfactory sensation is instrumental for perceiving risks in many everyday practices (Hsu 2020). Yet an encounter with potentially virus-laden aerosol clouds has a different quality. Perception has to deal with imperceptible elements. Here, our "all-too-human sensorium" (Myers 2020, 76) tends to fail at registering the aerosol that concerns us exactly in its withdrawal. All the more intensely felt is the uncertainty about the atmospheric interplay between human bodies, particles, viruses, and air currents. It is this highly indeterminate potential to infect and be infected that made atmospheres of COVID-19 so volatile. Dealing in her fieldwork with the H5N1 influenza virus, also spread via the aerosol pathway, anthropologist Celia Lowe aptly speaks of "cloudy insecurities" (Lowe 2010, 626). With the radical uncertainty about the atmospheric contact zones, the discomfort zones multiply. Especially the figure of the "healthy carrier" (see Wald 2008, 68–113) without symptoms is prone to materialize affectively as a kind of background suspicion that "tinctures" life in bodily proximity to others. In the extreme, this suspicion may grow into paranoia about the atmospheric infiltration of pathogenic mixtures, an affect that might be characterized in terms of an "elemental horror."

ATMOSOCIAL LIFE IN PERMEABLE CLOUDS

In his philosophy of elemental media, John Durham Peters (2015, 30) laconically claims that "ontology is not flat; it is . . . cloudy." With this statement, Peters distances himself from Bruno Latour's notion of a flat ontology of actor networks. By conceiving being as atmospheric, he implicitly opposes two-dimensional to three-dimensional spaces, clearcut edges to fuzzy boundaries, solid associations to turbulent fluids, the dry to the wet, and not least, connectivity between network nodes to sensitivity to the weather. This final section expands on this list of attributes that circumscribe the contours of the atmosocial condition that the COVID-19 pandemic has brought to collective experience. It will show that life in clouds forces us to attend to the permeability of bodies and environments. This permeability has important consequences for sociological concepts of relationality and relatedness.

During the controversy around aerosols, scientific articles and public communication often included visual representations that depict scenes of airborne transmission between humans. Sociologically these representations are diagrams that display a mode of relatedness particular to ecologies of breath—figure 11.1 is an example of this. It originated in a medical journal (Milton 2020, 414) and was used in a much-read FAQ article by a group of renowned atmospheric and environmental scientists who made a strong case for the significance of airborne infection (Jimenez et al. 2021). The picture depicts aerosols in three different sizes floating around interacting bodies. The "index case" on the left appears

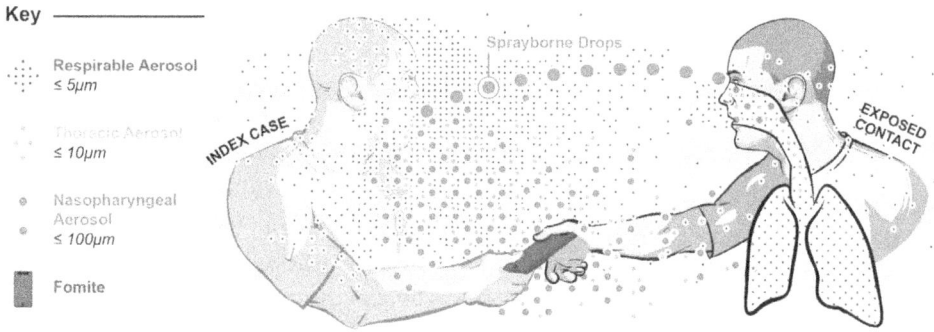

Key

- ∴∴ **Respirable Aerosol**
  ≤ 5μm

- **Thoracic Aerosol**
  ≤ 10μm

- ⊙ **Nasopharyngeal Aerosol**
  ≤ 100μm

- ▮ **Fomite**

Sprayborne Drops

INDEX CASE

EXPOSED CONTACT

FIG. 11.1. Transmission of droplet aerosols. Source: Milton (2020); also used in Jimenez et al. (2021, sect. 1.1).

as a "cloud body," a body "surrounded by a communicable mist" (Brown et al. 2020, 978). The body of the index case almost morphs into its bioaerosol emissions where the entity of dispersal becomes a dispersed entity. The "exposed contact" on the right inhales the aerosols. How deep the particles invade the respiratory tract depends on the particle size, the smallest particles reaching the lower airways. The contrast between the aerosol and the two other transmission pathways is instructive for understanding the atmosocial. Even if all three pathways imply the moment microbes leave and enter organisms, the aerosol component most strongly relativizes the idea of the body as a discrete entity. Bodies are marked by porous boundaries that render them vulnerable in ecologies of breath. COVID-19 has thus foregrounded an atmospheric "biointimacy" (Weston 2017, 71–104).

This intimacy of life in shared atmospheres goes against the classical sociological concept of interaction. In fact, both the transmission via "spray-borne" droplets and via a contaminated object are closely tied to interactive behaviors, either through droplets ballistically released while speaking or through touch. The volatility and aerial persistence of aerosols, in contrast, extends the atmosocial beyond interactive settings. The study mentioned above concerning the COVID-19 outbreak in a "poorly ventilated restaurant" (Li et al. 2021) is illustrative of this since the infected individuals across the different tables did not interact. Other portrayals of airborne transmission frequently depict scenes of people just standing next to each other in public transportation. In the atmosocial, intimacies of shared breath thus result from a type of relation that is in many cases "interpassive" (Seyfert 2019, 150–75), not interactive. The atmosocial goes against the "activist bias in sociology" (Müller 2016, 157) that manifests itself in the disciplinary focus on performances and forms of "doing" (doing gender, doing age, doing culture, doing time, etc.). Activities between persons are not necessary to enter an atmosocial relation; mutual passivity suffices.

A look at another case study helps develop this point further: the investigation of an outbreak of nine confirmed COVID-19 cases in early 2020 that most likely took place via fecal aerosols in a high-rise building (Kang et al. 2020). The three infected families lived in vertically aligned flats. As in the case study of the restaurant, an experimental simulation was made using a tracer gas. It found evidence that "bioaerosols were generated during toilet flushing and then spread via the draining stacks and vents" (Kang et al. 2020, 979) across twelve floors. The study underscores the complexity of fluid dynamics occurring in the built environment. Differences in air pressure, the geometry of drainage infrastructures, wind speed, and the buoyancy effect in vent and stack pipes all played a role. These dynamics involved the residents in an atmosocial intimacy that bears traces of what probably ranges among the most abject encounters imaginable—an encounter mediated though the "bioaerosolization of wastewater, mixed with urine, feces, and exhaled mucus" (Kang et al. 2020, 979). Here, the uncanny appearance of the atmosocial is pushed to the extreme with the introduction of a radical otherness into the home, both repulsive and threatening. The atmosocial possesses "powers of horror" due to its relationship with abject materialities (Kristeva 1982).

However, one does not have to *do* much to engage in this affectively heavily laden mode of elemental interaction. According to the authors of the case study, the infected persons very likely never met face to face during the period under investigation. Their atmosocial encounter did not even take place in the form of an interpassive relation, as it might have occurred, for instance, in the elevator where people typically avoid social interaction. Bodies do not have to be with each other in a room at the same time to enter an atmosocial relation: it is enough to live "alongside," understood as a form of relationality that results from "dwelling amidst different kinds" (Latimer 2013, 81). The atmosocial therefore has the capacity to befall or invade those living alongside each other.

More generally, the case study highlights the intricate link between atmospheres and habitation. In her book *The Forgetting of Air in Martin Heidegger*, Luce Irigaray (1999) rhetorically asks, "Is not air the whole of our habitation as mortals? . . . Can man live elsewhere than in air?" (8). The virus-laden aerosol cloud is an elemental medium that is "part of the habitat" (Peters 2015, 4), especially when it comes to the inhabitation of indoor spaces. Its properties pose a challenge to what we have come to understand as hygiene. Traditionally, most hygiene practices seek to sanitize boundary zones to prevent the incursion of germs, both into bodies and into interior spaces more broadly. However, when it comes to aerosols, the dispensers with hand sanitizer at the entrances of shops, universities, hairdressers, or administrative buildings are quite useless. Experts on environmental health and air quality have made a case for considering building ventilation systems as relevant for infection control efforts (Morawska et al. 2021). They envision the atmosocial as a technologically highly mediated space and, as such, as a matter for engineering solutions. Whereas in the twentieth century air conditioning was predominantly used for comfort (Murphy 2006, 20–23), COVID-19 gave it a new purpose in aerial hygiene. In the pandemic ventilation rates and filtering capacities became key parameters for atmospheric safety. However, the purification of the atmospheric habitat has limits that derive from the aerial medium itself. Even those authors who most forcefully advocate for building ventilation systems to be included in the arsenal of public health technologies would concede that "air as a contagion medium is nebulous, widespread, not owned by anybody, and uncontained" (Morawska et al. 2021, 689).

The permeability of bodies by their aerial habitat creates a disease ecology that has similarities with those resulting from pesticide use in agriculture (see Nash 2007, 127–69) or from radiation (Weston 2017, 71–104). The difference lies in the active contribution that bodies make to their habitat. Bodies bathe in the atmospheres that they coproduce and are immersed in the emissions resulting from their conviviality. On the one hand, bodies are vital forces immanent to their milieu. They always already change the atmosphere on which they depend through their mere presence. As Emma Garnett has pointed out in her analysis of computer simulations of indoor airflows, "occupant bodies were shown to participate in the way buildings breathe" (Garnett 2020, 68). At the same time bodies are pervaded by their aerial milieu. Relations via gases, moisture, and fluids are not relationships between discrete entities. In this sense atmosocial relations differ from network relations. The image of the network typically presents clear-cut nodes connected through straight edges that are neatly drawn. Its topology abstracts from the elemental constitution of any given milieu. While the concept of entanglement has recently been introduced to question the notion of discrete entities, it still depicts relations as interwoven threads. Yet, in the atmosocial, bodies become enmeshed in rather than entangled with their surroundings. Life in cloudy ecologies of breath thus foregrounds some of the "non-human powers circulating around and within human bodies" (Bennett 2010, ix). Such powers emanate from speaking and working bodies, depend on their number and their behavior, their incubation time and their individual viral load, architectural and infrastructural designs, window openings and ventilation systems, temperature and humidity. All these relations impact and form atmosocial encounters.

THE ATMOSOCIAL IN THE TRADING ZONE

This article has focused on the problematization of airborne transmission in the context of the SARS-CoV-2 pandemic. The practical stakes are high. The controversy regarding transmission pathways not only harbors important implications for how to organize interactive and interpassive relations to minimize infection risks, but also has begun to change the epistemic constitution of public health. Air as a medium has confronted public health institutions with forms of knowledge and expertise that for decades had only a marginal impact on them (Greenhalgh et al. 2021b). Importantly, atmospheric chemists, physicists, and building engineers have not only gained authority on the topic over time; they have also introduced concepts, instruments, and experimental procedures with new criteria on what counts as scientific evidence. These differ from those considered by microbiology and population epidemiology. Faced with the intricacies of what aerosol clouds can actually do in ecologies of breath, the very heterogeneous expert knowledges have entered into what the historian of science Peter Galison (2010) has called a "trading zone." Here the process of infection is reconceived as occurring in ecologies of breath. Accordingly, infection control now faces the problem of securing indoor air space.

However, it was not necessarily the goal of this chapter to point out the transformation of knowledge and its related governmental valence. Instead, the aim was to approach the debate over airborne transmission as an event that contains a conceptual lesson for sociology. The analysis approached a form of knowledge production far removed from the sociological orbit to understand the societal aggregate of "the atmosocial," which has gained practical virulence since early 2020. Following the airborne pathway, three arguments were made: First, the atmosocial has a voluminous shape. It transcends territorial organizations of space and subverts their promise of order. Even if attempts at territorializing atmospheres via sound, smell, or architectural design are widespread in many areas of contemporary life (Philippopoulos-Mihalopoulos 2016), the fluid medium cannot be bound. Compared with "territories of the self" as understood by Goffman (1971), atmospheres of the self are difficult to precisely delimit and localize. Second, the atmosocial has a turbulent state. The fluid dynamics of respiratory life are integral to the affective dynamic of highly uncertain atmospheric encounters and can only approximately be calculated. Breath and affect are irreducibly tied to each other. Together, they account for the speculative quality of the atmosocial. Third, the atmosocial neither exhibits the structure of an interaction nor that of a network, but rather has a cloudy constitution. Instead of a world of discrete relations with demarcated boundaries, it highlights the cohabitation of permeable bodies in built environments. Due to atmospheric properties of those environments, it may suffice to live alongside each other to establish atmosocial contact.

Unfolding these atmosocial implications enshrined in the controversy over airborne transmission is a prerequisite for the social sciences and the humanities for engaging in discussions about the impact of breathing and air conditioning on health in the future. Although these debates are, at their very core, about collective life, contributions from sociology have been scarce. Since the atmosocial challenges disciplinary assumptions about what constitutes a social fact, deep-seated conceptual settings are at least partly responsible for this lack of engagement. In this chapter it was therefore necessary to depart from established modes of theorizing. Prominently Richard Swedberg (2016) has stressed the importance of imaginative practices for adequately capturing novel phenomena. He seems to consider analogical forms to be fit for the purpose. His recommendation is to "look . . . for a structural similarity between something that is well understood and the phenomenon you are studying" (Swedberg 2016, 11). Yet the atmosocial can hardly be effectively approached in this way. This chapter did not generate sociological access to a sociologically relevant, but surprisingly neglected, phenomenon by using established disciplinary means. It rather sought to stay within what Isabel Stengers (2021, 72), quoting Clive Barnet, has termed the "situational provocation of the present"—in this case, the situational provocation contained in the proposition of airborne transmission of microbes. The aim was not to support or discard this proposition but to inquire into how airborne transmission brings to the fore relations that make a difference for collective being, as well as the modes by which they are perceived and accounted for. For this purpose, works from physics, atmospheric science, and engineering were read, to some extent, as social theory. The concept of the atmosocial is the result of this process. It is the path through which sociology may itself enter the "trading zone," to test if and how it may participate in addressing a key challenge of the present.

ECOLOGIES, DISEASES, AND SECURITY

EMBRACING CHANGES?

Melanie A. Kiechle

John Mansfield was touring Europe in 1871. So was cholera. The prospect that the two might meet weighed heavily on Mansfield's father James, who repeatedly sent John health advice in lengthy letters from New York. James wrote about the reports of cholera's ravages on August 20: "We [your mother and I] hope you will locate near the mountains or in that portion of the country where that dreaded disease has not made its appearance" (Mansfield 1871b). Though James and Mary Mansfield were neither physicians nor scientists, they thought about diseases in relation to place and topography. When John contemplated traveling to Rome for his art studies, his parents urged him to wait until "it is cooler, after the city becomes free from sickness" (Mansfield 1871a). The Mansfields' advice to their son boiled down to a deceptively simple task: "Look out and steer clear of any epidemic that seems to be contagious" (Mansfield 1871a). For the Mansfields, "health security" was neither all encompassing nor provided by governments, but an individual goal that required vigilance of surrounding environs.

Though written a century and a half ago, James Mansfield's advice to John strikes a contemporary chord because of its recent echoes in the COVID-19 pandemic. As public health in the United States conformed to a culture that celebrates individualism and capitalism, authorities replaced mandates for shared behavior and funding for public health measures with the suggestion that individuals should protect themselves from the airborne virus (Lincoln and Sosin 2022). The "you do you" approach to public health frustrated many in a moment when epidemiologists and physicians considered informed risk assessment around coronavirus transmission as nigh impossible (Jetelina 2022). As a result, Americans were left to unwittingly follow Mansfield's 1871 advice and avoid spaces where the epidemic was known to spread.

As the introduction to this volume explains and Mansfield's advice exemplifies, thinking about disease spatially and ecologically is not new. Neither is the goal of health security. Yet the contours of thinking about epidemic ecologies and the urgency of disease control have sharpened in the late twentieth and early twenty-first centuries, as disease eradication faltered in the face of emerging infectious diseases such as HIV/AIDS and SARS. In line with other discussions of "preparedness"—for terrorism, for natural disasters, for manifold crises—national governments and international organizations have prioritized preparation for the next disease outbreak. Recent efforts at disease preparedness have taken on ecological and more-than-human dimensions, with close attention paid to imperiled ecosystems and places where pathogens are likely to cross from nonhuman to human populations. Much as recent advice echoes the Mansfields' 1871 guidance, these modern efforts have multiple antecedents.

The previous chapters explore attempts to understand and prevent diseases across two and a half centuries. Reading these cases together reveals how much and how little has changed. This concluding commentary attends to three major points of congruence within ecological approaches to health security: disease terminology and the creation of authority, existent practices and emerging technologies, and social relations. It then turns attention to some of the persistent difficulties for ecological thinking with a consideration of individual lives and the long-term impacts of disease events.

TERMINOLOGY AND EXPERTISE

What is an illness? What constitutes an epidemic? While these questions seem straightforward, the cases surveyed here provide a wide range of answers. In every consideration of a disease outbreak, both the naming of a new disease and the decisions about how to respond are events

Melanie A. Kiechle

that reveal existing power dynamics. These dynamics in turn influence responses to the disease and its impact on society.

In the cases of pellagra and yaws, both diseases that primarily afflicted laboring populations (see Karina Turmann, chap. 2; and Julia Engelschalt, chap. 3), health authorities defined disease often by ignoring or dismissing the experiences and knowledge of those who endured these illnesses. The physicians who strove to explain pellagra in the early twentieth century ignored obvious contributing factors such as poverty and malnutrition in order to fixate on theories that fit their preconceived notions about environment and illness. When yaws struck Caribbean plantations in the eighteenth century, physicians and plantation managers adopted the disease name from those afflicted but ignored yaws sufferers' other knowledge and methods. Rather than recognize enslaved women's efforts to protect their children, or acknowledge that poor living conditions contributed to the spread of disease, authorities asserted that yaws was a product of Black women's behavior. This interpretation conformed to the prevailing discourse among whites on both sides of the Atlantic about racial difference and enabled disease responses that strengthened the system of enslavement.

Place has also played a role in the construction of expertise. Ironically, proximity to a disease outbreak has not always conveyed authority. Within colonial societies, authorities often opined from afar, explaining diseases through their beliefs about environments and constitutions rather than through direct observation. When physicians and public health officials have been in situ for disease outbreaks, their conflicting loyalties have led to multiple actions. As Susan Jones highlights, the Soviet scientists and physicians who studied disease environments in Central Asia were both representatives of the state and committed to helping local populations (chap. 1). Even while they furthered collectivization, Soviet scientists spread sanitary science and offered treatments that both aided Kazakh residents and established these newcomers as authorities whom locals could trust. This trust in turn furthered the goals of the Soviet state, especially when Soviet scientists enlisted women and children in the destruction of their landscape.

The "armchair epidemiologists" of the early twentieth century eschewed place altogether in their disease models, remaining in Britain while they aimed to explain disease throughout the empire. As Lukas Engelmann explains, these epidemiologists were statisticians committed to mathematical modeling (chap. 5). Because previous epidemic curves were so regular, these epidemiologists looked for a universal law and ignored factors that interfered with universality, including both human constitutions and local environments. Such thinking worked only until influenza virulently emerged in the 1918–1919 pandemic, spurring renewed interest in factors once favored in disease causation, including human constitutions.

EXISTING PRACTICES AND AVAILABLE TECHNOLOGIES

Who has secured epidemic ecologies? What tools have they used to accomplish this? Despite constant innovations and new technologies, there have been many continuities in practices of disease control since 1850. Perhaps the most significant continuity is in health authorities, who seemingly have not changed even as technologies for observation and intervention have evolved. Henning Füller's examination of syndromic surveillance, an innovation in data collection employed in the United States since 2001, reveals that this new effort to collect and monitor diverse data streams has not better detected disease outbreaks or changed the response to novel diseases (chap. 6). On the contrary, the epidemiologists and public health officials who first worked with syndromic surveillance insisted that the system is not sensitive enough for early detection. These officials expected that physicians would continue to raise the first alarm about a new disease, after which authorities could use the data collected to investigate patterns. Because the authorities over health remain the same—physicians for humans, veterinarians for nonhumans—syndromic surveillance has not been able to overcome the ontological divisions between authorities immersed in different fields with different terminologies. Furthermore, the political investment in syndromic surveillance superseded its actual utility to those in the field of public health. Rather than further investment in syndromic surveillance resulting from the system's efficacy at identifying emerging threats, political authorities in the 2010s planned to expand syndromic surveillance because of their belief that more data equals better understanding.

Throughout the COVID-19 pandemic, there has been significant discussion of society's "new normal," a phrase that refers to previously uncommon behaviors and patterns of thought that had become widespread. Much of this conversation has been driven by cultural commentators who marveled at the difficulty of defining the new normal in labor practices, daily interactions, and social norms. However, as the chapters by Henning Füller and Steve Hinchliffe argue, data collection is constantly redefining normal with little examination or comment. Hinchliffe notes that the data collection within livestock farming does not function in ways that protect the health of current herds or flocks, but records events that might be helpful for the health of future flocks (chap. 7). The constant monitoring of farm and animal conditions neither cares for existing flocks and herds nor improves the contemporary farm labor,

Melanie A. Kiechle

but operates to obscure the tacit knowledge of farmworkers and intensify farms' production for markets. Data collection is a market necessity and thus increasingly market driven. Both syndromic surveillance and livestock diagnostics are examples of "shifting baseline syndrome" (Pauly 1995, 430), in which the constant accumulation of data creates new standards as much—if not more than—it identifies emerging problems. While these practices build on the assumption that constant surveillance will quickly identify new disease events, these surveillance systems miss gradual, ongoing changes—like the compromising of immune systems outlined by Hannah Landecker in chapter 10—that make populations susceptible to disease. In these examples, securing epidemic ecologies is oriented toward future unknowns rather than addressing contemporary health problems.

Of course, any "new normal" is based on and defined against the previous normal—this is one of the reasons that extending the consideration of securing epidemic ecologies before the 1980s is so valuable. Both racial beliefs and beliefs that certain environments were inherently healthy or sickly were the normal that shaped research programs and actions in the past. For the USSR in the early twentieth century, securing the epidemic ecology of the steppe was a radical program of total environmental control premised on the belief that the environment itself harbored disease. As Soviet researchers identified the many disease vectors of the steppe, the solution to protect health was the destruction of every plant, animal, and habitat that might harbor disease (Jones, chap. 1). In contrast to total environmental control, other societies have tried to secure health through total human control. Caribbean authorities combatted yaws by imprisoning Black mothers and separating these women from their children (Turmann, chap. 2). This action was based on existing racism; rather than creating a "new normal," it deepened the social divisions that already existed and hardened beliefs that health differed by race.

SOCIAL RELATIONS

In the final chapter of this volume, Sven Opitz argues that the recognition of an airborne virus requires a reconceptualization of social relations (chap. 11). Rather than consider social relations as a product or cocreation of shared spaces and communal organizations, Opitz suggests shifting attention to the shared atmosphere. The atmosocial that Opitz outlines is more capacious than social interactions and thus transcends—or perhaps binds—long-standing social divisions by class, race, ethnicity, religion, and nation-state. There are no barriers in the atmosocial; instead, when we consider airborne pathogens and the role of air pollution in disease susceptibility, people are bound across rooms, buildings, and the globe by the air that they share. Recognizing the atmosocial requires understanding that bodies are permeable, in ways that were more common under Hippocratic medicine (Valencius 2002), and reconceptualizing notions of intimacy. The conceptualization of the atmosocial offers some optimism for securing health. When extended to its fullest, recognizing the intimacy and bodily entanglements of sharing air could unify the globe—or it might, as public health expert Charles V. Chapin pessimistically opined about aerial disease transmission in 1912, be so overwhelming as to discourage any preventive action.

Just as the social is changed now, so is the body. Ecological thinking, as it relates to securing ecologies and protecting health, must reconsider the body's interiors as well as the surrounding landscape. Hannah Landecker's chapter introduces the myriad changes that have been documented at epithelial barriers and emerged as the "preconditions" for severe cases of COVID-19 (chap. 10). Landecker explains that in bodies frequently exposed to irritants, the immune system is always in a heightened state of activity before it encounters a new virus or pathogen. Thus previous understandings of the immune system, that it "attacks" invading pathogens and "secures" the body, are no longer operational. If we continue to think of the immune system in these dated terms of bodily defense, all immune systems are compromised by fending off the continual attacks of environmental exposures. Reconceptualizing the human body as an anthropogenic biome necessitates also reconceptualizing health security. Securing epidemic ecologies in the Anthropocene requires not only vigilance against zoonotic diseases that cross into human populations, but close attention to how the human-changed environment is in turn changing human bodies.

Discussions of security are always discussions of potential violence, but the violence of emerging diseases and health threats is already here. In the favelas on urban peripheries, violence is as constant as the ongoing environmental exposure that have made bodies susceptible to emerging diseases. As Oswaldo Santos Baquero, Sara Cristina Aparecida da Silva, and Júlia Amorim Faria argue in chapter 4, this violence is neither accidental nor inconsistent with health emergencies. Instead, the violence contributes to the health emergency and poses nonviral threats to human health. In already violent conditions such as these, securing epidemic ecologies is not enough to protect health and lives. Much as the conceptualizations of the atmosocial and anthropogenic biome call for new definitions and operations of health security, countering the impact of everyday violence on health emergencies entails working to create a "new normal" in which everyday violence is a thing of the past.

INDIVIDUALS AND THE HARMS THAT LINGER

In January 1867 Celestia A. Webster noted a change in her body. In her diary, Webster recorded, "My throat troubles me considerable this winter. I begin to be afraid it is going to be something serious" (Webster 1867, January 9). Webster's fears were realized the following week, when a respiratory illness forced her from the school where she taught to her boardinghouse bed and then, a month later, back to her familial home. With a lung that "did not perform its office," as a physician informed her, Webster could no longer teach and struggled to complete daily tasks (Webster 1867, January 26). Celestia Webster's bout with illness upended her life, not for a few days or weeks, but for her remaining fourteen years.

Framing health security through epidemic ecologies illuminates much, but misses the experiences of individuals like Webster. Infectious diseases affect the masses, but individuals experience and react to illness on their own terms. Webster, like so many individuals around the world, closely observed and felt changes in her body, but remained unaware of the ecology she shared with the unknown virus that sickened her. Could she have secured her body and health within such an ecology? Unlike Mansfield, whose travels might expose him to disease, Webster fell ill in her home environment. What could have been done by the state—from its smallest level of the family and household to its largest incarnations at the nation and the globe—to ensure health?

Webster's encounter with illness, from which she neither recovered nor died, raises an important question about the harms that linger. Recognizing that state and global organizations play a leading role in securing epidemic ecologies, what role do they take in caring for lives harmed by illness? Many of the chapters in this volume have focused on efforts to stop the advance of disease and prevent death, but death is not the only outcome of infectious disease. As Ann H. Kelly and Clare Herrick document in their chapter about the Ebola outbreak in West Africa, surviving the outbreak is not the end of dealing with Ebola (chap. 8). Though the emergency R&D and application of ring vaccination succeeded in containing Ebola during the West African outbreak, this success was not repeated in the Congo because of the latter's large number of cases and limited resources. Furthermore, West African Ebola survivors encounter a strong social stigma, considered pariahs by their peers because of the potential that these survivors might be the source of the next outbreak. Conversely, survivors are highly valued by pharmaceutical companies conducting research and development. Survivors' bodies are useful for an emerging market, but survivors' lives are harmed for years after their encounter with the disease.

Like many issues covered in this volume, questions about the ethics of conducting medical research are not new. These questions arise whenever we consider how scientific and medical knowledge has been obtained through the study of others' suffering. This occurred in gynecology (Cooper Owens 2017), in the development of epidemiology through slavery and wars (Downs 2021), in the notorious Tuskegee syphilis experiments (Jones 1993, Reverby 2009), and in the ongoing life and study of Henrietta Lacks's cells without Lacks's knowledge or consent and long after her death (Skloot 2010). Given the profits that companies are making from emergency R&D—not to mention the societies, often in the Global North, spared disease outbreaks because of research conducted elsewhere—what is the responsibility of those with power to care for the lives harmed by disease?

Issues of responsibility loom large as the definition of COVID-19 as an epidemic emergency concludes, but the disease persists. As many scholars of disability have insisted, it is important to recognize that the COVID-19 pandemic is both a mass death event and a mass disabling event (National Academies of Sciences, Engineering, and Medicine 2022; Ely 2023). Those who survive their encounter with the SARS-CoV-2 virus are at risk for the lingering multiple effects labeled as "long COVID," a term created and legitimated by sufferers rather than medical practitioners (Callard and Perego 2021). Perhaps traditional medical authorities will take control of defining and treating long COVID; when this happened in the cases of yaws (Turmann, chap. 2) and pellagra (Engelschalt, chap. 3), authorities ignored the knowledge and practices of those with the diseases. Alternatively, long COVID sufferers as activists and research participants might shape both medical research and the definition of expertise, as occurred in HIV/AIDS (Epstein 1995). The simultaneous commodification of Ebola survivors' bodies for research and lack of care for Ebola survivors' well-being in society is a recent and cautionary tale (Kelly and Herrick, chap. 8). At the time of writing, those suffering from long COVID are still the foremost authorities on the condition—and increasingly neglected in public discussions and cognizance of COVID-19's impacts on contemporary society (Wu 2023; Yong 2023).

One of the challenges for considering lingering harms such as long Covid is inherent in thinking about health in terms of security. As discussed in the introduction, security is the framework for approaching emerging diseases because of the ways in which bioterrorism and disease outbreaks threaten nation-states: their stability, their power structures, their populace. Just as widespread disease has the potential to disrupt social order, widespread disabling alters a society. Celestia Webster lived

Melanie A. Kiechle

for fourteen years with long-term health impairments from her respiratory ailment. Unable to teach or to perform sustained household labor, Webster relied on her family and her local community for care and support. Webster's life after illness is both a success of communal caregiving and a portrait of disability.

Disability is most often discussed in terms of labor and productivity, building on the long-standing definition of public health—both what it is and its value for society—through worker productivity and economic benefits (Rosen 1974; LeRoux 2016a, 2016b; Olivarius 2019; Kiechle 2021) rather than around quality of life. Yet diminished quality of life and the increased need for caregiving are disease outcomes that alter social order and can disrupt societies. Following Hannah Landecker's analysis of the Anthropocene within contemporary bodies (chap. 10), how many people will suffer as Celestia Webster did? Are these lives being protected and valued or, as in the case of the livestock farms that Hinchliffe (chap. 7) studies, are these casualties of market forces and better data for "next time"? The coincidence of the Anthropocene and COVID-19 is syndemic, much like the coincidence of violence in favelas and the COVID-19 pandemic (Baquero, da Silva, and Faria, chap. 4). Yet these crises are not mere coincidence. As Eleanora Rohland (2020) has suggested, these likely are branches of the same crisis—a situation that calls for an even more capacious discussion of security than the one begun in this volume.

Responsibility for health, like responsibility for the global climate changes of the late Anthropocene, is not universal. While health outcomes are borne by many, the responsibility follows lines of power, in both political and economic terms. Securing epidemic ecologies needs to be defined not only against the crisis of encountering an emerging disease, but also to encompass the long-term impacts of these diseases on bodies and societies.

NOTES

CHAPTER 1: SECURING THE STEPPES

1. Party leadership structures changed often during the period covered by this chapter, but the General Secretariat, the CCCP (Central Committee), and the Politburo were architects and executors of policy in Central Asia, with the support of periodic Communist Party Congresses.

2. The classic introduction to early Soviet collectivization, in English, is Davies (1980).

3. Today's independent nations of Turkmenistan, Kyrgyzstan, Tajikistan, and Kazakhstan are vibrant republics located in a large continental landmass between China and Russia. The "'Stans" have long and often contentious histories with colonial powers (Russia and Britain) and with their neighbors. Their borders, landscapes, and the ethnic makeup of their peoples have changed dramatically over the past two hundred years. For convenience and brevity, this chapter uses today's names for these very dynamic regions. To increase accessibility for those who can read English but not Russian, the use of Russian words will be minimized, and sources will be provided in English translation, if available.

4. For the best discussion of "genocide" versus "crimes against humanity" as applied to the 1930s Kazakh steppe, see Cameron (2018, 6, 15, 177–79). In grim detail, Cameron's book documents the devastating transition for nomadic Kazakhs forcibly integrated into the Stalinist drive for ever-increasing crop production.

5. An excellent set of essays on the 1916 Central Asian uprising is Chokobaeva, Drieu, and Morrison (2020).

6. The Soviet People's Commissariat of Public Health, Narkomzdrav, changed names in 1945 to Minzdrav (Ministry of Public Health).

7. Important recent works in Soviet environmental history that foreground nonhuman historical actors include Bruno (2016); Kraikovski and Lajus (2019); Peterson (2019); and Moon (2013).

8. With some important exceptions, including Michaels (2003).

9. Today's Kazakh steppes were historically incorporated into the Soviet Union in stages: in 1867–1918, it was part of Russian Turkestan; in 1925, most of it became part of the Kazakh ASSR; and in 1936, it became the Kazakh Soviet Socialist Republic (SSR).

10. Khudyakov and Suchkov (1999)'s article was included in a series of newsletters self-published by retired plagueologist Moisey Iosevitch Levi and circulated among his colleagues. Most copies are in private hands in Russia. One complete copy, collected by Raymond A. Zilinskas and colleagues at the Center for Nonproliferation Studies (Monterey, California), exists in the United States, held by the Russian and Commonwealth Independent State Collection, Hoover Institution, Stanford University, CA.

11. For example, the major Moscow Institute for Experimental Medicine, named for N. G. Gamaleya, had a very active Division of Natural Focal Infections, led by Pavlovsky and subsequently by Petrishcheva during this period, and this approach remains important in Soviet epidemiology and experimental medicine today (Korenberg 2018).

12. Historian Loren Graham argues that an epigenetic version of Lysenko's ideas has gained ascendency again in Vladimir Putin's Russia (L. Graham 2016).

13. The literature on Soviet science is vast, but the classic introductions in English are Graham (1994) and Krementsov (1997).

14. For a brief biography of Petrishcheva, see Kryuchechnikov (2000).

15. Pavlovsky's important early articles include Pavlovsky (1939, 1946). He published a mature statement of the theory in Pavlovsky (1966 [1965]). For a biography of Pavlovsky, see Prokhorova (1972).

16. For detailed remembrances of the scientists working with Pavlovsky, see Shtakelberg (1930).

17. Pavlovsky 1928–1932. "Kirghiz" at this time referred to ethnic Kazakh and Kirghiz peoples. Today's independent republic is spelled "Kyrgyzstan."

18. For example, see Polina Andreevna Petrishcheva to Pavlovsky, Telegrams and correspondence, 1931, Fond 878, Opus 4, no. 189, l. 17, ARAN-P.

19. For the history of Tajik delimitation and incorporation into the USSR, see Kassymbekova (2016).

20. This information was not published or relayed (as required) to the World Health Organization. Politburo decrees, renewed at multiple Party congresses, made outbreaks of plague a state secret or for "internal use only." This secrecy devolved from the fact that the USSR had officially declared its territory to be free of plague, and from concerns that the extent of the Soviet biological weapons program (which used plague microorganisms) would be discovered by the West. These data were compiled from unpublished scientific accounts in Levi (1991–1999, 2:3–26, 5:214–16).

21. See, for example, Shaw, "Mastering Nature"; Oldfield and Shaw, *The Development of Russian Environmental Thought*; and Brain, "The Great Stalin Plan for the Transformation of Nature."

22. My thanks to Andrea Wiegeshoff for drawing my attention to this point.

CHAPTER 2: PLANTATIONS AND "THE YAWS"

I sincerely thank the editors of this book and organizers of the conference "Securing Epidemic Ecologies: Spaces of Disease Control in Historical Perspective," and all contributors to the conference, for their constructive feedback and many helpful comments, and Joana Amaral and Stephen Foose for their diligent proofreading.

1. *Treponema pallidum pertenue* is a subspecies of the *Treponema* genus. *Treponema pallidum endemicum* is the causative agent of endemic or non-venereal syphilis or bejel, whereas *Treponema carateum* causes pinta. The species *Treponema pallidum pallidum* is the causative agent of venereal syphilis. Its *Treponema* origin was microscopically detected for the first time in 1905 by Fritz Schaudinn and Erich Hoffmann. There are also numerous nonpathogenic *Treponema* species in the oral cavity, the genital region, and the intestine (Bundesministerium für Gesundheit 2002, 818).

2. Today, yaws is diagnosed via a TPPA (*Treponema pallidum* particle agglutination assay) and RPR (rapid plasma reagin test) in the serum, a PCR (polymerase chain reaction) from a biopsy or egg aspiration of the lesion as well as immunological rapid tests that are, however, still in development (Schneider 2021).

3. The interaction between colonialism and ideas about race is extensively studied within the range of the recent medical histories: e.g., Hogarth (2017); Handler and Bilby (2012); Schiebinger (2010, 2013); Browne (2017); Lowe, Payne-Jackson, and Johnson (2009).

4. On the subject of amelioration, see Spence (2014).

5. When British legislation banned the slave trade on paper in 1807, the hope was that an adequate supply of labor would regulate itself through re-

production. However, the slave registers recorded a decline in the labor force, which fueled discussion about promoting maternity among enslaved women. "Rewarding" mothers was a common strategy within the plantation system. Managament instructions recommended giving certain amounts of money on special occasions like Christmas but also when women gave birth, and bonuses when their children survived. As many newborns died within the first thirty days, these bonuses were presumably given after a certain period of time (see Lascelles et al. 1786, 31).

6. When *Black* and *white* as dichotomous terminology are used here, it depicts historical Eurocentric ascriptions that generally referred to colonized Africans or people of African descent and colonizing Europeans, and is obviously a simplification considering the wide range of ethnic and social diversity and heterogeneity in the region in focus. Moreover, the use of the terms *African*, *black*, and *enslaved* became blurred during this period or were often used congruently.

7. Until the nineteenth century the basic approaches of humoral pathology were still part of the medical standard, meaning that "aspects of outward appearance, might be altered by long residence in a different climate. In other words, both internal and external features were regarded as amenable to change." In the early nineteenth century these views changed and the theory of a physiological adaption without a change of outward characteristics was dominant. Based on this logic, diseases and their treatments were seen to "differ according to racial type, suggesting fixed anatomical differences." Complete acclimatization was generally considered impossible. However, it was conceivable that races alien to the climatic environment could approximately acclimate themselves by adapting certain lifestyles, such as in diet or clothing (Harrison 2010, 113–14).

8. The topic of midwifery during slavery is huge and deserves more attention than is possible in this chapter; due to the limited space I refer to the elaborate work on this topic by Tara Inniss (2009).

9. On the experience of maternal grief of enslaved mothers see, e.g., Turner (2017).

CHAPTER 3: "A DISEASE OF PLACE"

This contribution is based on research funded by the Collaborative Research Center "Practices of Comparing" at Bielefeld University, Germany. The author would like to thank Susan Jones and the other participants of the "Securing Epidemic Ecologies" workshop for their feedback on an earlier version of this text.

1. Scholars tend to agree that microbial etiologies of pellagra were embraced by (particularly southern) political actors because they indirectly exonerated poverty (e.g., Beardsley 1987; Crabb 1992; Bollet 1992).

2. This terminology (in particular, the term *dementia*) may seem somewhat dated, but is still used in contemporary medical literature (Cao, Wang, and Cestodio 2020).

3. The phrase "benevolent assimilation" originates from President William McKinley's "Benevolent Assimilation Proclamation," issued on December 21, 1898 (Miller 1982).

CHAPTER 5: MODELS, EPIDEMIC THEORY, AND THE DECLINE OF THE ENVIRONMENT SINCE THE 1900S

1. This is partly due to the equivalent development of formal and mathematical approaches in postwar ecology. In the 1980s modeling came close to achieving the status of a universal science, where questions of ecology, economy, and social dynamics converged with the climate, physics, and budding data science. See, e.g., Morgan 2012; Edwards 2010; Anderson 2021.

2. Warwick Anderson has argued for the case of Australia in the same period that the idea of associating the environment with disease had become a nostalgic view, when the conviction grew that all environmental constraints can be overcome to settle white populations on the Australian continent and that there are no pathological boundaries affixed to the material world. Pathologies and their reservoirs, in other words, were to be found only in the dynamics of populations (Anderson 2000).

3. It is of course important to acknowledge that Ross eventually sought to develop a theory of epidemics beyond infectious diseases and that his "theory of happenings" was indeed meant to discern laws of the spread of anything that could potentially be seen to distribute across social structures.

4. According to Fine, these efforts could be identified as a tradition of modeling that then continues into the late twentieth century. Neither an a priori pathometry nor an empirical fitting of curves to historical data, modeling is described as a feedback loop between theory and data (Fine 1975).

CHAPTER 7: THE DIAGNOSTIC MACHINE

The research and writing of this chapter were made possible as a result of the Diagnostics and Animal Livestock project funded by the Economic and Social Research Council (Grant No. ES/P008194/1).

1. The extracts are taken from a research project on the use of diagnostic devices in livestock farming and work that involved a survey of veterinary surgeons, as well long interviews and focus groups with farmers, veterinarians, and test developers.

CHAPTER 8: CONTACTS AND CONTACTS OF CONTACTS: EBOLA RINGS, RESERVOIRS, AND FRONTIERS

1. Richard Hatchett, interview by Stephen Morrison, *Take as Directed* (podcast), August 6, 2019, https://soundcloud.com/csis-57169780/battling -against-pathogens.

2. Hatchett interview.

3. Tedros Adhanom Ghebreyesus, speech at High-Level Event on Ebola Virus Disease in DRC, July 15, 2019, https://www.who.int/director-general/speeches/detail/high-level-event-on-ebola-virus-disease-in-drc.

4. Natalie Roberts, quoted in "Independent Ebola Vaccination Committee Is Needed to Overcome Lack of WHO Transparency," MSF press release, September 23, 2019, https://www.msf.org/ebola-drc-independent-ebola-vaccination-committee-needed.

5. See Dionne and Seasay 2016, "American Perceptions of Africa during an Ebola Outbreak" in *Ebola's Message*, eds. Nicholas G. Evans, Tara C. Smith, and Maimuna S. Majumder (Cambridge, MA: MIT Press), retrieved from https://covid-19.mitpress.mit.edu/pub/ialzqbqw.

6. This is particularly acute under conditions of global emergency, where states are often required to institute policies that go against their own interests for the sake of global containment.

CHAPTER 9: LAZILY INVASIVE?

This publication draws on research that receives generous funding from the Volkswagen Foundation within the interdisciplinary project "Mobile Mosquitoes—Understanding the Entangled Mobilities of *Aedes* Mosquitoes and Humans in India, Mexico, Tanzania and Germany" (Grant Number 9B366). Uli Beisel and Carsten Wergin would like to further acknowledge their international partners and partner institutions in this project, in particular our fellow coinvestigators Fredros Okumu (Ifakara Health Research Institute, Tanzania), Gerardo Suzán (National Autonomous University of Mexico), Ashwani Kumar (Indian Council of Medical Research), and Norbert Becker (Institut für Dipterologie, Germany). Further information can be found via: http://portal.volkswagenstiftung.de/search/projectDetails.do?ref=9B366.

We also thank Sven Opitz, Carolin Mezes, and Andrea Wiegeshoff for their kind editorial support, careful reading, and their patience.

CHAPTER 10: THE UNDERLYING CONDITION AS A DISEASE ECOLOGY

1. Space does not allow for teasing apart all the different conjectures about how diabetes, obesity, atherosclerosis, and nonalcoholic fatty liver disease have been considered separately as risk factors for severe outcomes from SARS-CoV-2 infection, or together, in terms of an umbrella effect of poor blood glucose regulation or other physiological states shared across these different metabolic diseases. Theories about the intersections of obesity and COVID include mechanical theories such as increased weight on the lungs making respiratory distress more likely, and tissue susceptibility theories such as the idea that adipose cells are a target for viral infection because of an increased number of target ACE2

receptors in that cell type. At the time of this writing, it remains controversial as to whether pancreatic tissue is a particular target for viral infection. A good overview of the theories and the evidence for each can be found in Drucker 2021.

2. Obviously we immediately move here to socially formatted exposures that follow and generate structures of health inequality that are also these molecular landscapes. See Kenner 2018; and Ahmann and Kenner 2020.

3. The "innate" or "nonspecific" immune system consists of the various barriers and protections against infection that are not specific to any one pathogen, such as antimicrobial peptides, clotting factors, signals that initiate inflammatory responses, and cells such as macrophages that phagocytose microbes and release signals that summon other immune cells to the site of the injury. Dendritic cells are part of the innate immune system that present antigens to the antibody-producing cells of the adaptive immune system.

4. I am agnostic as to whether one would prefer *capitalocene, plantationocene, cthulucene,* or whatever one likes here in place of Anthropocene (Haraway 2015). Anthropogenically generated, biologically consequent features of industrialization are not significant because they are made ubiquitous and homogenous by large-scale production and dissemination and are therefore everywhere the same for everyone, but rather the opposite: these historically specific features of mass production and biological impact are characteristic in that they manifest as uneven effects everywhere, enhancing our responsibility to understand them in the landscape of the cell as part of the unequal legacies of industrial production and extraction (Liboiron 2021).

5. Chronic inflammation in metabolic disorders has also been linked to the presence of increased adipose (fat) cell release of inflammatory signals into the body. While I am focused here on explanatory frameworks that concern the ecology of the mucosal boundary, these are not the only ways that chronic inflammation is understood and studied.

6. For a critical discussion of the concept of cellular exhaustion, see Kaminski, Lemoine, and Pradeu 2021.

7. Timothy Lenton, Sébastien Dutreuil, and Bruno Latour (2020) have argued that earth systems scientists and biologists have treated life as the property shared by living beings, impacted by and contained in the abiotic planetary systems under study. This perspective, they argue, has precluded a Gaian perspective on "Life" as all living beings on Earth descended from a single common ancestor, biochemically and behaviorally engaged in generating the habitability of its own environment, and thus the habitability of the biosphere. They argue for a perspective that instead reasons "from organisms' metabolisms outwards," and redefines the Anthropocene as, in part, Life's impact on Earth, not just an abiotic environment that is or is not habitable for living things (Lenton, Dutreuil, and Latour 2020, 248).

CHAPTER 11: THE ATMOSOCIAL CONDITION

1. My translation.

2. At the time of writing in winter 2022–2023, it remains to be seen which elements of the pandemic interaction order will persist. A somewhat anecdotal but nonetheless illustrative example of the softening of COVID-19 regulations and its perception as an endemic disease was the distribution of three stickers to the participants at the conference of the Annual Anthropological Association 2022 in Seattle: a red sticker that read "6 feet apart—No Contact No Exceptions," a yellow sticker "Elbow Bump!" and a green sticker "High Fives & Handshakes Plus Frequent Hand Washing." The attendees were encouraged to visibly attach these stickers to their conference badge. In this way, uncertainties about preferences could be ruled out, and hence the interaction order was rendered explicit in a personalized manner.

3. The institute is named after Hermann Rietschel who held the world's first chair for ventilation and heating at Königlich Technische Hochschule zu Berlin between 1883 and 1910. For further information, see https://www.hri.tu-berlin.de/menue/home/hermann_rietschel/.

4. A particle as large as ten micrometers can not only "stay in the air for minutes, and thus not fall to the ground in a few seconds. At typical indoor speeds of 0.1 meter/second, a 10 μm aerosol can travel approx. 50 meters" (Jimenez et al. 2021, sect. 2.6.).

5. Laboratories in which research on highly pathogenic organisms is conducted use negative pressure or "airlock" security doors to control aerosol movements. As Melanie Armstrong (2017, 68–96) has shown, the development of biological weapons and high-security laboratories have a common genealogy. Both are connected by a "fear of air" (82). One may add, admittedly randomly, that the album "Fear of Music" (1979) by the band Talking Heads features a song called "Air" containing the verse: "Where is the protection that I needed? Air can hurt you too." The relationship between the atmospheric and the sonic suggested by the album title would open a wholly new register. For a poignant account of the environmental imaginary at work in the idea of air being "polluted" by noise, see Peterson (2022, 77–107).

6. The notion of "gehauchte Kommunen" in the German original is even more evocative. A literal translation would be "aspirated communes."

7. For instance, the German weekly newspaper *Die Zeit* created an "interactive tool [that] shows how the coronavirus spreads" to "find out how safe your environment is" (Dinklage et al. 2020).

REFERENCES

Abramowitz, Sharon. 2017. "Epidemics (Especially Ebola)." *Annual Review of Anthropology* 46 (1): 421–45.

"An Abstract of the Meliorating Provisions of the Laws Last Enacted in Each of the British West India Colonies for the Government and Protection of the Slave Population." 1831. GD142/57/3, National Records of Scotland, Edinburgh.

Adaylton, Franco, José Santos, Vinícius Konchinski, Fernanda Canofre, Matheus Rocha, and Paulo Dias. 2022. "Defensorias veem aumentar casos de furto de comida na pandemia." Folha de São Paulo. https://www1.folha.uol.com.br/cotidiano/2022/03/defensorias-veem-aumentar-casos-de-furto-de-comida-na-pandemia.shtml.

Adey, Peter, Kevin Hannam, Mimi Sheller, and David Tyfield. 2021. "Pandemic (Im)mobilities." *Mobilities* 16 (1): 1–19. doi:10.1080/17450101.2021.1872871.

Ahmann, Chloe, and Alison Kenner. 2020. "Breathing Late Industrialism." *Engaging Science, Technology, and Society* 6: 416–38. doi:10.17351/ests2020.673.

Akdis, Cezmi A. 2022. "The Epithelial Barrier Hypothesis Proposes a Comprehensive Understanding of the Origins of Allergic and Other Chronic Noncommunicable Diseases." *Journal of Allergy and Clinical Immunology* 149 (1): 41–44. doi:10.1016/j.jaci.2021.11.010.

Alagona, Peter, Jane Carruthers, Hao Chen, Michèle Dagenais, Sandro Dutra e Silva, Gerard Fitzgerald, Shen Hou, et al. 2020. "Reflections: Environmental History in the Era of COVID-19." *Environmental History* 25 (4): 595–686. doi:10.1093/envhis/emaa053.

Alvim, Angélica Tanus Benatti, Volia Regina Costa Kato, and Jeane Rombi de Godoy Rosin. 2015. "A urgência das águas: Intervenções urbanas em áreas de mananciais." *Cadernos Metrópole* 17 (33): 83–107. doi:10.1590/2236-9996.2015-3304.

Amanzholova, Dina. 2004. "Kazakh Autonomy: From the Concept of Nations to Soviet Self-Determination." *Acta Slavica Iaponica* 21: 115–43.

Amanzholova, Dina. 2009. *Na izlome: Alash v etnopoliticheskoi istorii Kazakhstana* [At the break: The Alash in the ethnopolitical history of Kazakhstan]. Almaty, KZ: Taymas.

Amato, Laura, Maria G. Dente, Paolo Calistri, Silvia Declich, and on behalf of the MediLabSecure Working Group. 2020. "Integrated Early Warning Surveillance: Achilles' Heel of One Health?" *Microorganisms* 8 (1): 1–10. doi:10.3390/microorganisms8010084.

Amorim, Júlia Faria, Sara Cristina Aparecida da Silva, Ana Claudia Camargo Gonçalves Germani, Dora Mariela Salcedo Barrientos, Gislene Aparecida Santos, and Oswaldo Santos Baquero. 2021. "Vivências da pandemia na São Remo: Um discurso coletivo." São Paulo: Instituto de Estudos Avançados—Universidade de São Paulo. doi:10.5281/zenodo.5211531.

Amsterdamska, Olga. 2004. "Achieving Disbelief: Thought Styles, Microbial Variation, and American and British Epidemiology, 1900–1940." *Studies in History and Philosophy of Science Part C: Studies in History and Philosophy of Biological and Biomedical Sciences* 35 (3): 483–507. doi:10.1016/j.shpsc.2004.06.001.

Amuasi, John H., Tamara Lucas, Richard Horton, and Andrea S. Winkler. 2020. "Reconnecting for Our Future: The Lancet One Health Commission." *The Lancet* 395 (10235): 1469–71. doi:10.1016/s0140-6736(20)31027-8.

Anderson, Ben. 2017. *Encountering Affect: Capacities, Apparatuses, Conditions.* Abingdon, UK: Routledge.

Anderson, Ben, and James Ash. 2015. "Atmospheric Methods." In *Nonrepresentational Methodologies. Re-envisioning Research*, edited by Phillip Vannini, 44–61. Abingdon, UK: Routledge.

Anderson, Warwick. 2000. "Geography, Race and Nation: Remapping 'Tropical' Australia, 1890–1930." *Medical History Supplement* 44 (S20): 146–59.

Anderson, Warwick. 2004. "Natural Histories of Infectious Disease: Ecological Vision in Twentieth-Century Biomedical Science." *Osiris* 19: 39–61. doi:10.1086/649393.

Anderson, Warwick. 2006. *Colonial Pathologies: American Tropical Medicine, Race, and Hygiene in the Philippines*. Durham, NC: Duke University.

Anderson, Warwick. 2010. "From Plague, an Epidemic Comes: Recounting Disease as Contamination and Configuration." In *Plagues and Epidemics: Infected Spaces Past and Present*, edited by D. Ann Herring, and Alan Swedlund, 251–68. Oxford: Berg.

Anderson, Warwick. 2021. "The Model Crisis, or How to Have Critical Promiscuity in the Time of Covid-19." *Social Studies of Science* 51 (2): 167–88. doi:10.1177/0306312721996053.

Araujo, Carla. 2001. "As marcas da violência na constituição da identidade de jovens da periferia." *Educação e Pesquisa* 27 (1): 141–60. doi:10.1590/S1517-97022001000100010.

Armelagos, George J., Kathleen C. Barnes, and James Lin. 1996. "Disease in Human Evolution: The Re-emergence of Infectious Disease in the Third Epidemiological Transition." *National Museum of Natural History Bulletin for Teachers* 18: 1–6.

Armstrong, Melanie. 2017. *Germ Wars: The Politics of Microbes and America's Landscape of Fear*. Oakland: University of California Press.

Arnold, David, ed. 1998. *Imperial Medicine and Indigenous Societies*. MSI ed. Studies in Imperialism. Manchester: Manchester University Press.

Asokan, Vaithinathan G., Ramanathan K. Kasimanickam, and Vanitha Asokan. 2013. "Surveillance, Response Systems, and Evidence Updates on Emerging Zoonoses: The Role of One Health." *Infection Ecology & Epidemiology* 3 (1): 21386. doi:10.3402/iee.v3i0.21386.

Bancil, Aaron S., Alicia M. Sandall, Megan Rossi, Benoit Chassaing, James O. Lindsay, and Kevin Whelan. 2021. "Food Additive Emulsifiers and Their Impact on Gut Microbiome, Permeability, and Inflammation: Mechanistic Insights in Inflammatory Bowel Disease." *Journal of Crohn's and Colitis* 15 (6): 1068–79. doi:10.1093/ecco-jcc/jjaa254.

Baquero, Oswaldo Santos. 2021a. "Ambiente e animais sinantrópicos." In *Censo vizinhança USP: Características domiciliares e socioculturais do Jardim São Remo e Sem Terra*, edited by Eliana Sousa Silva, Érica Peçanha, and Dalcio Marinho Gonçalves, 158–60. São Paulo: Instituto de Estudos Avançados da Universidade de São Paulo.

Baquero, Oswaldo Santos. 2021b. "Ambiente e animais sinantrópicos." In *Censo vizinhança USP: Características domiciliares e socioculturais do Jardim Keralux e da Vila Guaraciaba*, edited by Eliana Sousa Silva, Érica Peçanha, and Dalcio Marinho Gonçalves, 158–60. São Paulo: Instituto de Estudos Avançados da Universidade de São Paulo.

Baquero, Oswaldo Santos. 2021c. "Animais de companhia." In *Censo vizinhança*

USP: Características domiciliares e socioculturais do Jardim São Remo e Sem Terra, edited by Eliana Sousa Silva, Érica Peçanha, and Dalcio Marinho Gonçalves, 153–57. São Paulo: Instituto de Estudos Avançados da Universidade de São Paulo.

Baquero, Oswaldo Santos. 2021d. "Animais de companhia." In Censo vizinhança USP: Características domiciliares e socioculturais do Jardim Keralux e da Vila Guaraciaba, edited by Eliana Sousa Silva, Érica Peçanha, and Dalcio Marinho Gonçalves, 153–57. São Paulo: Instituto de Estudos Avançados da Universidade de São Paulo.

Baquero, Oswaldo Santos. 2021e. "Cultivo de plantas no domicílio." In Censo vizinhança USP: Características domiciliares e socioculturais do Jardim Keralux e da Vila Guaraciaba, edited by Eliana Sousa Silva, Érica Peçanha, and Dalcio Marinho Gonçalves, 160. São Paulo: Instituto de Estudos Avançados da Universidade de São Paulo.

Baquero, Oswaldo Santos. 2021f. "Cultivo de plantas no domicílio." In Censo vizinhança USP: Características domiciliares e socioculturais do Jardim São Remo e Sem Terra, edited by Eliana Sousa Silva, Érica Peçanha, and Dalcio Marinho Gonçalves, 160. São Paulo: Instituto de Estudos Avançados da Universidade de São Paulo.

Baquero, Oswaldo Santos. 2021g. "One Health of Peripheries: Biopolitics, Social Determination, and Field of Praxis." Frontiers in Public Health 9: 1–12. doi:10.3389/fpubh.2021.617003.

Baquero, Oswaldo Santos, Mario Nestor Benavidez Fernández, and Myriam Acero Aguilar. 2021. "From Modern Planetary Health to Decolonial Promotion of One Health of Peripheries." Frontiers in Public Health 9: 1–11. doi:10.3389/fpubh.2021.637897.

Baraniuk, Chris. 2021. "Covid-19: What Do We Know about Airborne Transmission of SARS-CoV-2?" British Medical Journal 373: n1030.

Bard, Alison M., David Main, Emma Roe, Anne Haase, Helen Rebecca Whay, and Kristen K. Reyher. 2019. "To Change or Not to Change? Veterinarian and Farmer Perceptions of Relational Factors Influencing the Enactment of Veterinary Advice on Dairy Farms in the United Kingdom." Journal of Dairy Science 102 (11): 10379–94. doi:10.3168/jds.2019-16364.

Bardosh, Kevin, Melissa Leach, and Annie Wilkinson. 2016. "The Limits of Rapid Response: Ebola and Structural Violence in West Africa." In One Health, edited by Kevin Bardosh, 260. London: Routledge.

Barker, Kezia. 2015. "Biosecurity: Securing Circulations from the Microbe to the Macrocosm." Geographical Journal 181 (4): 357–65. doi:10.1111/geoj.12097.

Barnes, Stephen A. 2011. Death and Redemption: The Gulag and the Shaping of Soviet Society. Princeton, NJ: Princeton University Press.

Bar-On, Yinon M., Rob Phillips, and Ron Milo. 2018. "The Biomass Distribution on Earth." *Proceedings of the National Academy of Sciences* 115 (25): 6506–11. doi:10.1073/pnas.1711842115.

Barrett, Frank A. 2000a. "August Hirsch: As Critic of, and Contributor to, Geographical Medicine and Medical Geography." *Medical History Supplement* 20: 98–117.

Barrett, Frank A. 2000b. *Disease and Geography: The History of an Idea.* Vol. 23. York, UK: York University Press.

Barron, Emma, Chirag Bakhai, Partha Kar, Andy Weaver, Dominique Bradley, Hassan Ismail, Peter Knighton, et al. 2020. "Associations of Type 1 and Type 2 Diabetes with COVID-19-Related Mortality in England: A Whole-Population Study." *The Lancet. Diabetes & Endocrinology* 8 (10): 813–22. doi:10.1016/S2213-8587(20)30272-2.

Bartlett, Harriet, Mark A. Holmes, Silviu O. Petrovan, David R. Williams, James L. N. Wood, and Andrew Balmford. 2022. "Understanding the Relative Risks of Zoonosis Emergence under Contrasting Approaches to Meeting Livestock Product Demand." *Royal Society Open Science* 9 (6): 211573. doi:10.1098/rsos.211573.

Bartumeus, Frederic, Guilherme B. Costa, Roger Eritja, Ann H. Kelly, Marceline Finda, Javier Lezaun, Fredros Okumu, et al. 2019. "Sustainable Innovation in Vector Control Requires Strong Partnerships with Communities." *PLoS Neglected Tropical Diseases* 13 (4): e0007204. doi.:10.1371/journal.pntd.0007204.

Bashford, Alison, and Sarah W. Tracy. 2012. "Introduction: Modern Airs, Waters, and Places." *Bulletin of the History of Medicine* 86 (4): 495–514. doi:10.1353/bhm.2012.0084.

Beardsley, Edward H. 1987. *A History of Neglect: Health Care for Blacks and Mill Workers in the Twentieth-Century South.* Knoxville: University of Tennessee Press.

Becker, Norbert, Björn Pluskota, Achim Kaiser, and Francis Schaffner. 2012. "Exotic Mosquitoes Conquer the World." In *Arthropods as Vectors of Emerging Diseases*, edited by Heinz Mehlhorn, 31–60. Berlin: Springer.

Becker, Norbert, Stefanie Schön, Alexandra-Maria Klein, Ina Ferstl, Ali Kizgin, Egbert Tannich, Carola Kuhn, Björn Pluskota, and Artur Jöst. 2017. "First Mass Development of *Aedes albopictus* (Diptera: Culicidae)— Its Surveillance and Control in Germany." *Parasitology Research* 116: 847–58. doi:10.1007/s00436-016-5356-z.

Beinart, William, and Lotte Hughes. 2007. *Environment and Empire.* Oxford: Oxford University Press.

Beisel, Uli, and Carsten Wergin. 2021. "Understanding Multispecies Mobilities: From Mosquito Eradication to Coexistence." In *Mosquitopia: The Place of Pests in a Healthy World*, edited by Marcus Hall and Dan Tamir, 33–46. London: Routledge.

Bennett, Jane. 2010. *Vibrant Matter: A Political Ecology of Things*. Durham, NC: Duke University Press.

Benöhr, Jens, Maike Brinksma, Ross Donihue, David Farò, Antonia Lara, Kara Lena Virik, Alejandro Ponce de León, et al. 2022. "Ecopolitical Mapping: A Multispecies Research Methodology for Environmental Communication." *Ciencias Sociales* 36: 317–43. doi:10.18046/recs.i36.5275.

ben Ouagrham-Gormley, Sonia. 2006. "Growth of the Anti-plague System during the Soviet Period." *Critical Reviews in Microbiology* 32: 33–46.

ben Ouagrham-Gormley, Sonia, Alexander Melikishvili, and Raymond Zilinskas. 2002 "The Soviet Anti-Plague System in the Newly Independent States, 1992 and Onwards." CNS Occasional Paper 18. Monterey, CA: James Martin Center for Nonproliferation Studies.

Benson, Etienne. 2020. *Surroundings. A History of Environments and Environmentalisms*. Chicago: University of Chicago Press.

Benton, Adia. 2017. "Whose Security? Militarization and Securitization during West Africa's Ebola Outbreak." In *The Politics of Fear: Médecins Sans Frontières and the West African Ebola Epidemic*, edited by Michiel Hofman and Sokhieng Au, 25–50. Oxford: Oxford University Press.

Berg, Marc. 1992. "The Construction of Medical Disposals: Medical Sociology and Medical Problem Solving in Clinical Practice." *Sociology of Health & Illness* 14 (2): 151–80. doi:10.1111/j.1467-9566.1992.tb00119.x.

Berlant, Lauren. 2007. "Slow Death (Sovereignty, Obesity, Lateral Agency)." *Critical Inquiry* 33 (4): 754–80. doi:10.1086/521568.

Bermudi, Patricia Marques Moralejo, Camila Lorenz, Breno Souza de Aguiar, Marcelo Antunes Failla, Ligia Vizeu Barrozo, and Francisco Chiaravalloti-Neto. 2021. "Spatiotemporal Ecological Study of COVID-19 Mortality in the City of São Paulo, Brazil: Shifting of the High Mortality Risk from Areas with the Best to Those with the Worst Socio-Economic Conditions." *Travel Medicine and Infectious Disease* 39: 101945. doi:10.1016/J.TMAID.2020.101945.

Berridge, Virginia. 2016. *Public Health: A Very Short Introduction*. Oxford: Oxford University Press.

Bhatia, Rajesh. 2020. "Need for Integrated Surveillance at Human-Animal Interface for Rapid Detection & Response to Emerging Coronavirus Infections Using One Health Approach." *Indian Journal of Medical Research* 151 (2–3): 132–35. doi:10.4103/ijmr.ijmr_623_20.

Birn, Anne-Emanuelle. 2014. "Backstage: The Relationship between the Rockefeller Foundation and the World Health Organization, Part I: 1940s–1960s." *Public Health* 128 (2): 129–40.

Blaxter, Mildred. 1978. "Diagnosis as Category and Process: The Case of Alcoholism." *Social Science & Medicine. Part A: Medical Psychology & Medical Sociology* 12: 9–17. doi:10.1016/0271-7123(78)90017-2.

Block, Kristen. 2017. "Slavery and Inter-imperial Leprosy Discourse in the Atlantic World." *Atlantic Studies* 14 (2): 243–62. doi:10.1080/14788810.2017
.1283474.

Blok, Anders, Moe Nakazora, and Brit Ross Winthereik. 2016. "Infrastructuring Environments." *Science as Culture* 25 (1): 1–22. doi:10.1080/09505431
.2015.1081500.

Böhme, Gernot. 1993. "Atmosphere as the Fundamental Concept of a New
Aesthetics." *Thesis Eleven* 36 (1): 113–26.

Bollet, Alfred Jay. 1992. "Politics and Pellagra: The Epidemic of Pellagra in the
U.S. in the Early Twentieth Century." *Yale Journal of Biology and Medicine*
65 (3): 211–21.

Bondurant, Eugene D. 1909. "Pellagra, with Report of Nine Cases." *Medical
Record (1866–1922)* 76 (8) (August 21): 300–304.

Bonwitt, Jesse, Michael Dawson, Martin Kandeh, Rashid Ansumana, Foday
Sahr, Hannah Brown, and Ann H. Kelly. 2018. "Unintended Consequenc-
es of the 'Bushmeat Ban' in West Africa during the 2013–2016 Ebola Virus
Disease Epidemic." *Social Science & Medicine* 200: 166–73.

Bordier, Arthur. 1884. *La géographie médicale.* Paris: C. Reinwald.

Boucquey, Noëlle, Kevin St. Martin, Luke Fairbanks, Lisa M. Campbell, and
Sarah Wise. 2019. "Ocean Data Portals: Performing a New Infrastructure
for Ocean Governance." *Environment and Planning D: Society and Space* 37
(3): 484–503. doi:10.1177/0263775818822829.

Bourouiba, Lydia. 2021. "The Fluid Dynamics of Disease Transmission." *Annual Review of Fluid Mechanics* 53 (1): 473–508.

Bowe, Emily, Erin Simmons, and Shannon Mattern. 2020. "Learning from
Lines: Critical COVID Data Visualizations and the Quarantine Quotidian." *Big Data & Society* 7 (2): 2053951720939236.

Bowker, Geoffrey, and Susan Leigh Star. 1999. *Sorting Things Out.* Cambridge,
MA: MIT Press.

Brada, Betsey. 2011. "'Not Here': Making the Spaces and Subjects of 'Global
Health' in Botswana." *Culture, Medicine, and Psychiatry* 35 (2): 285–312.

Brain, Stephen. 2010. "The Great Stalin Plan for the Transformation of Nature."
Environmental History 15 (4): 670–700.

Brathwaite, John. 1788. "Evidence Given before the Lords of the Privy Council." *Journal of the Barbados Museum and Historical Society* 1951 (18): 24–38.
Bridgetown: Barbados Museum and Historical Society.

Brende, Børge, Jeremy Farrar, Diane Gashumba, Carlos Moedas, Trevor Mun-
del, Yasuhisa Shiozaki, Harsh Vardhan, Johanna Wanka, and John-Arne
Røttingen. 2017. "CEPI—A New Global R&D Organisation for Epidemic
Preparedness and Response." *The Lancet* 389 (10066): 233–35.

Brighenti, Andrea Mubi. 2010. "On Territorology: Towards a General Science
of Territory." *Theory, Culture & Society* 27 (1): 52–72.

Broemer, R. 2000. "The First Global Map of the Distribution of Human Diseases: Friedrich Schnurrer's 'Charte über die geographische Ausbreitung der Krankheiten' (1827)." *Medical History Supplement* 44 (S20): 176–85.

Brown, Hannah, and Ann H. Kelly. 2014. "Material Proximities and Hotspots: Toward an Anthropology of Viral Hemorrhagic Fevers." *Medical Anthropology Quarterly* 28 (2): 280–303.

Brown, Nik, Chrissy Buse, Alan Lewis, Daryl Martin, and Sarah Nettleton. 2020. "Air Care: An 'Aerography' of Breath, Buildings and Bugs in the Cystic Fibrosis Clinic." *Sociology of Health & Illness* 42 (5): 972–86.

Brown, Nik, Chrissy Buse, Alan Lewis, Daryl Martin, and Sarah Nettleton. 2021. "The Coughing Body: Etiquettes, Techniques, Sonographies and Spaces." *Biosocieties* 16 (2): 270–88.

Brown, Tim. 2011. "'Vulnerability Is Universal': Considering the Place of 'Security'and 'Vulnerability' within Contemporary Global Health Discourse." *Social Science & Medicine* 72 (3): 319–26.

Browne, Randy M. 2017. *Surviving Slavery in the British Caribbean. Early American Studies.* Philadelphia: University of Pennsylvania Press.

Brownlee, John. 1906. "Statistical Studies in Immunity: The Theory of an Epidemic." *Proceedings of the Royal Society of Edinburgh* 26 (1): 484–521. doi:10.1017/S037016460002472X.

Brownlee, John. 1909. "Certain Considerations on the Causation and Course of Epidemics." *Proceedings of the Royal Society of Medicine* 2: 243–58.

Bruce, Ann, Katherine E. Adam, Henry Buller, Kin Wing Chan, and Joyce Tait. 2021. "Creating an Innovation Ecosystem for Rapid Diagnostic Tests for Livestock to Support Sustainable Antibiotic Use." *Technology Analysis & Strategic Management* 34 (11): 1–14. doi:10.1080/09537325.2021.1950678.

Bruno, Andy. 2016. *The Nature of Soviet Power: An Arctic Environmental History.* New York: Cambridge University Press.

Bryan, Charles S., and Shane R. Mull. 2015. "Pellagra Pre-Goldberger: Rupert Blue, Fleming Sandwith, and the 'Vitamine Hypothesis.'" *Transactions of the American Clinical and Climatological Association* 126: 20–45.

Bueno, Samira, David Marques, and Dennis Pacheco. 2021. "As mortes decorrentes de intervenção policial no Brasil em 2020." *Anuário Brasileiro de Segurança Pública* 15: 59–69.

Buller, Henry, Katie Adam, Alison Bard, Ann Bruce, Kin Wing Chan, Stephen Hinchliffe, Lisa Morgans, Gwen Rees, and Kristen K. Reyher. 2020. "Veterinary Diagnostic Practice and the Use of Rapid Tests in Antimicrobial Stewardship on UK Livestock Farms." *Frontiers in Veterinary Science* 7: 765. doi:10.3389/fvets.2020.569545.

Bundesministerium für Gesundheit. 2002. "Bekanntmachung des Arbeitskrei-

ses Blut des Bundesministeriums für Gesundheit. Treponema Pallidum."
Bundesgesundheitsblatt—Gesundheitsforschung—Gesundheitsschutz 45 (10): 818–26. doi:10.1007/s00103-002-0473-5.

Burnard, Trevor, and Richard Follett. 2012. "Caribbean Slavery, British Anti-slavery, and the Cultural Politics of Venereal Disease." *Historical Journal* 55 (2): 427–51. doi:10.1017/S0018246X11000513.

Bush, Vannevar. 2020 (1945). "Science, the Endless Frontier." In *Science, the Endless Frontier*. Princeton, NJ: Princeton University Press.

Butler, Judith. 2020. *The Force of Nonviolence: An Ethico-political Bind*. London: Verso.

Buttimer, Anne. 2000. "Airs, Waters, Places: Perennial Puzzles of Health and Environment." *Medical History* 44 (20): 211–16. doi:10.1017/S0025727300 07335X.

Buzan, Barry, Ole Wæver, and Jaap De Wilde. 1998. *Security: A New Framework for Analysis*. Boulder, CO: Lynne Rienner.

Byrd, W. Michael, and Linda A. Clayton. 2000. *An American Health Dilemma: A Medical History of African Americans and the Problem of Race, Beginnings to 1900*. New York: Routledge.

Caines, Clement. 1801. *Letters on the Cultivation of the Otaheite Cane. The Manufacture of Sugar and Rum; the Saving of Melasses; the Care and Preservation of Stock; with the Attention and Anxiety Which Is due to Negroes*. London: W. Bulmer.

Callard, Felicity, and Elisa Perego. 2021. "How and Why Patients Made Long Covid." *Social Science & Medicine* 268. https://doi.org/10.1016/j.socscimed.2020.113426.

Cameron, Sarah. 2018. *The Hungry Steppe: Famine, Violence, and the Making of Soviet Kazakhstan*. Ithaca, NY: Cornell University Press.

Cantor, David. 2002. *Reinventing Hippocrates*. Aldershot, UK: Ashgate.

Cao, Shanjin, Xiaodan Wang, and Kristen Cestodio. 2020. "Pellagra, an Almost-Forgotten Differential Diagnosis of Chronic Diarrhea: More Prevalent Than We Think." *Nutrition in Clinical Practice* 35 (5): 860–63.

Carlson, Colin J., Gregory F. Albery, and Alexandra Phelan. 2021. "Preparing International Cooperation on Pandemic Prevention for the Anthropocene." *BMJ Global Health* 6 (3): e004254. doi:10.1136/ bmjgh-2020–004254.

Cassidy, Angela. 2019. *Vermin, Victims and Disease: British Debates over Bovine Tuberculosis and Badgers*. Cham, Switzerland: Palgrave Macmillan.

CDC (Centers for Disease Control and Prevention). 2017. "Sustaining Global Health Security Is Critical to Protecting America's National Security." http://medbox.iiab.me/modules/en-cdc/www.cdc.gov/media/dpk/cdc-24-7/global-health-security-eid-supplement/index.html.

Chade, Jamil. 2022. "ONU denuncia 'alta exponencial' de violência policial durante Covid no país." UOL (Universo Online), February 14. https://noticias.uol.com.br/colunas/jamil-chade/2022/02/14/onu-cobra-do-brasil-respostas-de-combate-a-violencia-policial-no-pais.htm.

Chakrabarti, Pratik. 2014. *Medicine and Empire, 1600–1960.* Basingstoke, UK: Palgrave Macmillan.

Chan, Kin Wing, Alison M. Bard, Kathrine E. Adam, Gwen M. Rees, Lisa Morgans, Liz Cresswell, Stephen Hinchliffe, et al. 2020. "Diagnostics and the Challenge of Antimicrobial Resistance: A Survey of UK Livestock Veterinarians' Perceptions and Practices." *Veterinary Record* 187 (12): e125. doi: 10.1136/vr.105822.

Chan, Sarah, and John Harris. 2011. "Human Animals and Nonhuman Persons." *The Oxford Handbook of Animal Ethics*, 304–31. New York: Oxford University Press.

Chapin, Charles V. 1912. *The Sources and Modes of Infection.* 2nd, rev. and enlarged, ed. London: John Wiley and Sons.

Chassaing, Benoit, Tom Van de Wiele, Jana De Bodt, Massimo Marzorati, and Andrew T. Gewirtz. 2017. "Dietary Emulsifiers Directly Alter Human Microbiota Composition and Gene Expression Ex Vivo Potentiating Intestinal Inflammation." *Gut* 66 (8): 1414–27. doi:10.1136/gutjnl-2016-313099.

Chen, Hao. 2020. "Nonhuman Animals in a Human Pandemic: Past and Present." *Environmental History* 25: 604–7. doi:10.1093/envhis/emaa053.

Chokobaeva, Aminat, Cloé Drieu, and Alexander Morrison. 2020. *The Central Asian Uprising of 1916.* Manchester: Manchester University Press.

Cleary, Michelle, Deependra K. Thapa, Sancia West, Mark Westman, and Rachel Kornhaber. 2021. "Animal Abuse in the Context of Adult Intimate Partner Violence: A Systematic Review." *Aggression and Violent Behavior* 61: 101676. doi:10.1016/J.AVB.2021.101676.

Cohen, Jon. 2022. "Merck Locates Frozen Batch of Undisclosed Ebola Vaccine, Will Donate for Testing in Uganda's Outbreak." *Science*, October 23. https://www.science.org/content/article/uganda-may-use-destroyed-ebola-vaccine-merck-fight-its-growing-outbreak.

Cohn, Samuel, and Ruth Kutalek. 2016. "Historical Parallels, Ebola Virus Disease and Cholera: Understanding Community Distrust and Social Violence with Epidemics." *PLoS Currents* 8. doi:10.1371/currents.outbreaks.aa1f2b60e8d43939b43fbd93e1a63a94.

Coleman-Jones, Emma. 2010. "Ronald Ross and the Great Malaria Problem: Historical Reference in the Biological Sciences." *Journal of Biological Education* 33 (4): 181–84. doi:10.1080/00219266.1999.9655662.

Collier, Stephen, and Andrew Lakoff. 2021. *The Government of Emergency: Vital Systems, Expertise, and the Politics of Security.* Princeton: Princeton University Press.

Coole, Diana, and Samantha Frost. 2010. "Introducing the New Materialisms."

In *New Materialisms: Ontology, Agency, and Politics*, edited by Diana Coole and Samatha Frost, 1–43. Durham, NC: Duke University Press.

Cooper, Melinda. 2006. "Pre-empting Emergence: The Biological Turn in the War on Terror." *Theory, Culture & Society* 23 (4): 113–35. doi:10.1177/026 3276406065121.

Cooper Owens, Dierdre. 2017. *Medical Bondage: Race, Gender, and the Origins of American Gynecology*. Athens: University of Georgia Press.

Corvalán, Carlos, Simon Hales, and Anthony J. McMichael. 2005. *Ecosystems and Human Well-Being: Health Synthesis. A Report of the Millennium Ecosystem Assessment*. Geneva: WHO.

Crabb, Mary Katherine. 1992. "An Epidemic of Pride: Pellagra and the Culture of the American South." *Anthropologica* 34 (1): 89–103.

Craddock, Susan, and Steve Hinchliffe. 2015. "One World, One Health? Social Science Engagements with the One Health Agenda." *Social Science & Medicine* 129: 1–4.

Crampton, Jeremy W., and John Krygier. 2005. "An Introduction to Critical Cartography." *ACME: An International Journal for Critical Geographies* 4 (1): 11–33. https://acme-journal.org/index.php/acme/article/view/723.

Creager, Angela N. H. 2002. *The Life of a Virus: Tobacco Mosaic Virus as an Experimental Model, 1930–1965*. Chicago: University of Chicago Press.

Crookshank, Francis Graham. 1922. "First Principles of Epidemiology." In *Influenza: Essays by Several Authors*, edited by Crookshank, 11–30. London: William Heinemann.

Crosby, Alfred W. 1972. *The Columbian Exchange: Biological and Cultural Consequences of 1492*. Westport, CT: Greenwood.

Crozier, Alex, Selina Rajan, Iain Buchan, and Martin McKee. 2021. "Put to the Test: Use of Rapid Testing Technologies for Covid-19." *British Medical Journal* 372: n208. doi:10.1136/bmj.n208.

Cueto, Marcos. 1995. "The Cycles of Eradication: The Rockefeller Foundation and Latin American Public Health, 1918–1940." In *International Health Organisations and Movements 1918–1939*, edited by Paul Weindling, 222–43. Cambridge: Cambridge University Press.

Curtis, Luke. 2021. "$PM_{2.5}$, NO_2, Wildfires, and Other Environmental Exposures Are Linked to Higher Covid-19 Incidence, Severity, and Death Rates." *Environmental Science and Pollution Research* 28 (39): 54429–47. doi:10.1007/s11356-021-15556-0.

D'Abramo, Flavio, and Hannah Landecker. 2019. "Anthropocene in the Cell." *Technosphere Magazine*, March. https://technosphere-magazine.hkw.de/p/Anthropocene-in-the-Cell-fQjoLLgrE7jbXzLYr1TLNn.

D'Abramo, Flavio, and Sybille Neumeyer. 2020 "A Historical and Political Epistemology of Microbes." *Centaurus* 62 (2): 321–30. doi:10.1111/1600-0498.12300.

Daly, Jeanne. 2005. *Evidence-Based Medicine and the Search for a Science of Clinical Care*. Berkeley: University of California Press.

Damialis, Athanasios, Stefanie Gilles, Mikhail Sofiev, Viktoria Sofieva, Franziska Kolek, Daniela Bayr, Maria P. Plaza, et al. 2021. "Higher Airborne Pollen Concentrations Correlated with Increased SARS-CoV-2 Infection Rates, as Evidenced from 31 Countries across the Globe." *Proceedings of the National Academy of Sciences* 118 (12): e2019034118. doi:10.1073/pnas.2019034118.

Davies, R W. 1980. *The Industrialisation of Soviet Russia*. Vol. 1, *The Socialist Offensive: The Collectivisation of Soviet Agriculture, 1929–30*. London: Palgrave Macmillan.

Davies, Sara E. 2008. "Securitizing Infectious Disease." *International Affairs* 84 (2): 295–313.

Davies, Sara E. 2013. "National Security and Pandemics." *UN Chronicle* 50 (2): 20–24.

Davis, Alicia, and Jo Sharp. 2020. "Rethinking One Health: Emergent Human, Animal and Environmental Assemblages." *Social Science & Medicine* 258: 1–8. doi:10.1016/j.socscimed.2020.113093.

Davis, Angela. 2003. *Are Prisons Obsolete?* New York: Seven Stories Press.

de Almeida, João Rangel. 2015. "Epidemic Opportunities: Panic, Quarantines, and the 1851 International Sanitary Conference." In *Empires of Panic: Epidemics and Colonial Anxieties*, edited by Robert Peckham, 57–86. Hong Kong: Hong Kong University Press.

Declich, Siliva, and Anne O. Carter. 1994. "Public Health Surveillance: Historical Origin, Methods and Evaluation." *Bulletin of the World Health Organization* 72 (2): 285–304.

Delamou, Alexandre, Bienvenu Salim Camara, Jean Pe Kolie, Achille Diona Guemou, Nyankoye Yves Haba, Shannon Marquez, Abdoul Habib Beavogui, Therese Delvaux, and Johan van Griensven. 2017. "Profile and Reintegration Experience of Ebola Survivors in Guinea: A Cross-Sectional Study." *Tropical Medicine & International Health* 22 (3): 254–60.

Delamou, Alexandre, Thérèse Delvaux, Alison Marie El Ayadi, Abdoul Habib Beavogui, Junko Okumura, Wim Van Damme, and Vincent De Brouwere. 2017. "Public Health Impact of the 2014–2015 Ebola Outbreak in West Africa: Seizing Opportunities for the Future." *BMJ Global Health* 2 (2): e000202.

Denova-Gutiérrez, Edgar, Hugo Lopez-Gatell, Jose L. Alomia-Zegarra, Ruy López-Ridaura, Christian A. Zaragoza-Jimenez, Dwigth D. Dyer-Leal, Ricardo Cortés-Alcala, et al. 2020. "The Association of Obesity, Type 2 Diabetes, and Hypertension with Severe Coronavirus Disease 2019 on Admission among Mexican Patients." *Obesity* 28 (10): 1826–32. doi:10.1002/oby.22946.

Dent, William. 1819a. Dent to his mother, April 20. Typescript copies of 49 letters, 1808–1824, to his Parents, Brother and Cousin from Surgeon William Dent, re his Training in London and at Colchester Barracks, and Service with the 89th Regiment during the Peninsular War, 1810–1814, in Canada, 1814, with the Army of Occupation in France, 1815–1818, and in the West Indies, 1819–1824. RAMC/536, Royal Army Medical Corps Muniments Collection, Wellcome Collection, London.

Dent, William. 1819b. Dent to his cousin, August 12. Typescript copies of 49 letters, 1808–1824, to his Parents, Brother and Cousin from Surgeon William Dent, re his Training in London and at Colchester Barracks, and Service with the 89th Regiment during the Peninsular War, 1810–1814, in Canada, 1814, with the Army of Occupation in France, 1815–1818, and in the West Indies, 1819–1824. RAMC/536, Royal Army Medical Corps Muniments Collection, Wellcome Collection, London.

de Sousa, Ana Cristina Augusto. 2020. "O que esperar do novo marco do saneamento?" *Cadernos de Saúde Pública* 36 (12): e00224020. doi:10.1590/0102-311X00224020.

Deynze, Allen Van, Pablo Zamora, Pierre-Marc Delaux, Cristobal Heitmann, Dhileepkumar Jayaraman, Shanmugam Rajasekar, Danielle Graham, et al. 2018. "Nitrogen Fixation in a Landrace of Maize Is Supported by a Mucilage-Associated Diazotrophic Microbiota." *PLoS Biology* 16 (8): e2006352. doi:10.1371/journal.pbio.2006352.

Diener, Leander. 2021. "COVID-19 und seine Umwelt: Von einer Geschichte der Humanmedizin zu einer ökologischen Medizingeschichte?" *NTM Zeitschrift für Geschichte der Wissenschaften, Technik und Medizin* 29 (2): 203–11. doi:10.1007/s00048-021-00299-3.

Dietz, Klaus. 2009. "Epidemics: The Fitting of the First Dynamic Models to Data." *Journal of Contemporary Mathematical Analysis* 44 (2): 97–104. doi:10.3103/S1068362309020034.

Dillon, Michael, and Julien Reid. 2009. *The Liberal Way of War: Killing to Make Life Live.* London: Routledge.

Dinklage, Fabian, Annick Ehmann, Elena Erdmann, Moritz Klack, Maria Mast, Julian Stahnke, Julius Tröger, et al. 2020. "Why Is the Risk of Coronavirus Transmission So High Indoors?" *Die Zeit,* November 26. https://www.zeit.de/wissen/gesundheit/2020-11/coronavirus-aerosols-infection-risk-hotspot-interiors.

Dionne, Kim Yi, and Laura Seay. 2016. "American Perceptions of Africa during an Ebola Outbreak." In *Ebola's Message: Public Health and Medicine in the Twenty-First Century,* edited by Nicholas G. Evans, Tara C. Smith and Maimuna S. Majumder. Cambridge, MA: MIT Press. Retrieved from https://covid-19.mitpress.mit.edu/pub/ialzqbqw.

Donahue, Donald A. 2011. "BioWatch and the Brown Cap." *Journal of Home-*

land Security and Emergency Management 8 (1): 1–13. doi:10.2202/1547-7355.1823.

Downs, Jim. 2021. *Maladies of Empire: How Colonialism, Slavery, and War Transformed Medicine*. Cambridge, MA: Belknap Press of Harvard University Press.

Drake, Daniel. 1854. *A Systematic Treatise, Historical, Etiological, and Practical: On the Principal Diseases of the Interior Valley of North America, as They Appear in the Caucasian, African, Indian, and Esquimaux Varieties of Its Population.* 2nd ser. Philadelphia: Lippincott, Grambo.

Droumeva, Milena. 2017. "Soundmapping as Critical Cartography: Engaging Publics in Listening to the Environment." *Communication and the Public* 2 (4): 335–51. doi:10.1177/2057047317719469.

Drucker, Daniel J. 2021. "Diabetes, Obesity, Metabolism, and SARS-CoV-2 Infection: The End of the Beginning." *Cell Metabolism* 33 (3): 479–98. doi:10.1016/j.cmet.2021.01.016.

Duff, Cameron. 2014. *Assemblages of Health*. Dordrecht: Springer Netherlands. doi:10.1007/978-94-017-8893-9.

Dulius, Graziele Testa, Aline Winter Sudbrack, and Luiza Maria de Oliveira Braga Silveira. 2021. "Aumento da violência intrafamiliar e os fatores associados durante a pandemia de COVID-19: Revisão integrativa de literatura." *Saúde Em Redes* 7 (1suppl): 205–13. doi:10.18310/2446-4813 2021v7n1Sup.3381g613.

Dumit, Joseph. 2021. "Substance as Method: Bromine, for Example." In *Reactivating Elements: Chemistry, Ecology, Practice*, edited by Dimitris Papadopoulos, Maria Puig de la Bellacasa, and Natasha Myers, 84–107. Durham, NC: Duke University Press.

Dunk, James H., David S. Jones, Anthony Capon, and Warwick H. Anderson. 2019. "Human Health on an Ailing Planet: Historical Perspectives on our Future." *New England Journal of Medicine* 381 (8): 778–82.

Dunn, Richard. 2014. *A Tale of Two Plantations: Slave Life and Labor in Jamaica and Virginia*. Cambridge, MA: Harvard University Press.

Dürbeck, Gabriele. 2018. "Narrative des Anthropozän: Systematisierung eines interdisziplinären Diskurses." *Kulturwissenschaftliche Zeitschrift* 3 (1): 1–20. doi:10.25969/MEDIAREP/3602.

Durkheim, Émile. 1982 (1895). *The Rules of Sociological Method*. Edited with an Introduction by Steven Lukes. Translated by V. D. Halls. New York: Free Press.

Dyl, Joanna L. 2006. "The War on Rats versus the Right to Keep Chickens: Plague and the Paving of San Francisco, 1907–1908." In *The Nature of Cities: Culture, Landscape and Urban Space*, edited by Andrew C. Isenberg, 38–61. Rochester, NY: University of Rochester Press.

ECDC (European Centre for Disease Prevention and Control). 2018. *Towards One Health Preparedness*. ECDC Technical Report, Stockholm.

ECDC and European Food Safety Authority. 2022. "Mosquito Maps: Surveillance for Invasive Mosquitoes." Stockholm: ECDC. https://ecdc.europa.eu/en/disease-vectors/surveillance-and-disease-data/mosquito-maps.

Ecks, Stefan. 2020. "Multimorbidity, Polyiatrogenesis, and COVID-19." *Medical Anthropology Quarterly* 34 (4): 488–503. doi:10.1111/maq.12626.

Edwards, Paul N. 2010. *A Vast Machine: Computer Models, Climate Data, and the Politics of Global Warming*. Cambridge, MA: MIT Press.

Elbe, Stefan. 2009. *Virus Alert: Security, Governmentality, and the AIDS Pandemic*. New York: Columbia University Press.

Elbe, Stefan. 2010. *Security and Global Health: Towards the Medicalization of Insecurity*. Cambridge: Polity Press.

Elbe, Stefan, Anne Roemer-Mahler, and Christopher Long. 2015. "Medical Countermeasures for National Security: A New Government Role in the Pharmaceuticalization of Society." *Social Science & Medicine* 131: 263–71.

Elden, Stuart. 2013. "Secure the Volume: Vertical Geopolitics and the Depth of Power." *Political Geography* 34: 35–51.

Electronic Privacy Information Center. 2004. "Total 'Terrorism' Information Awareness (TIA)." https://epic.org/privacy/profiling/tia/.

Ellis, Erle C., Arthur H. W. Beusen, and Kees Klein Goldewijk. 2020. "Anthropogenic Biomes: 10,000 BCE to 2015 CE." *Land* 9 (5): 129. doi:10.3390/land9050129.

Elman, Cheryl, Robert A. McGuire, and Barbara Wittman. 2014. "Extending Public Health: The Rockefeller Sanitary Commission and Hookworm in the American South." *American Journal of Public Health* 104 (1): 47–58.

Ely, Wes. 2023. "As Covid Turns Three, Americans Play Disability Roulette." *Boston Globe*, May 26.

Engelmann, Lukas. 2018. "A Plague of Kinyounism: The Caricatures of Bacteriology in 1900 San Francisco." *Social History of Medicine* 33 (2): 489–514. doi:10.1093/shm/hky039.

Engelmann, Lukas. 2019. "Configurations of Plague: Spatial Diagrams in Early Epidemiology." *Social Analysis* 63 (4): 89–109. doi:10.3167/sa.2019.630405.

Engelmann, Lukas, Caroline Humphrey, and Christos Lynteris. 2019. "Introduction: Diagrams beyond Mere Tools." *Social Analysis* 63 (4): 1–19. doi:10.3167/sa.2019.630401.

Engelmann, Lukas, and Christos Lynteris. 2020. *Sulphuric Utopias: A History of Maritime Fumigation*. Cambridge, MA: MIT Press.

Engelmann, Lukas, Catherine M. Montgomery, Steve Sturdy, and Cristina Moreno Lozano. 2022. "Domesticating Models: On the Contingency of

Covid-19 Modelling in UK Media and Policy." *Social Studies of Science* 53 (1): 121–45. doi:10.1177/03063127221126166.

Engelmann, Sasha, and Derek McCormack. 2021. "Elemental Worlds: Specificities, Exposures, Alchemies." *Progress in Human Geography* 45 (6): 1402–18.

Enticott, Gareth. 2012. "The Local Universality of Veterinary Expertise and the Geography of Animal Disease." *Transactions of the Institute of British Geographers* 37: 75–88.

Epstein, Steven. 1995. "The Construction of Lay Expertise: AIDS Activism and the Forging of Credibility in the Reform of Clinical Trials." *Science, Technology, and Human Values* 20 .(4): 408–37.

Espinosa, Mariola. 2009. *Epidemic Invasions: Yellow Fever and the Limits of Cuban Independence, 1878–1930*. Chicago: University of Chicago Press.

Etheridge, Elizabeth W. 1972. *The Butterfly Caste: A Social History of Pellagra in the South*. Westport, CT: Greenwood.

Ettling, John. 1981. *The Germ of Laziness: Rockefeller Philanthropy and Public Health in the New South*. Cambridge, MA: Harvard University Press.

Evans, Nicholas. 2018. "Blaming the Rat? Accounting for Plague in Colonial Indian Medicine." *Medical Anthropology Theory* 5 (3): 15–42.

Everts, Jonathan. 2020. "The Dashboard Pandemic." *Dialogues in Human Geography* 10 (2): 260–64. doi:10.1177/2043820620935355.

Fairchild, Amy L., Ronald Bayer, and James Colgrove. 2007. *Searching Eyes: Privacy, the State, and Disease Surveillance in America*. Berkeley: University of California Press.

Fairhead, James, Melissa Leach, and Ian Scoones. 2012. "Green Grabbing: A New Appropriation of Nature?" *Journal of Peasant Studies* 39 (2): 237–61.

FAO (Food and Agriculture Organization). United Nations. 2020. "Build Back Better in a Post-COVID-19 World—Reducing Future Wildlife-Borne Spillover of Disease to Humans." Sustainable Wildlife Management (SWM) Programme. Rome. https://www.fao.org/3/cb1490en/CB1490EN.pdf.

Farley, John. 1991. *Bilharzia: A History of Imperial Tropical Medicine*. Cambridge: Cambridge University Press.

Farmer, Paul. 2016. "The Second Life of Sickness: On Structural Violence and Cultural Humility." *Human Organization* 75 (4): 279–88.

Farmer, Paul. 2001. *Infections and Inequalities: The Modern Plagues*. Berkeley: University of California Press.

Farmer, Paul. 2004. "An Anthropology of Structural Violence." *Current Anthropology* 45 (3): 305–25. doi.:10.1086/382250.

Farmer, Paul. 2020. *Fevers, Feuds, and Diamonds: Ebola and the Ravages of History*. New York: Farrar, Straus and Giroux.

Fattorini, Daniele, and Francesco Regoli. 2020. "Role of the Chronic Air Pollution Levels in the Covid-19 Outbreak Risk in Italy." *Environmental Pollution* 264: 114732. doi:10.1016/j.envpol.2020.114732.

Faver, Catherine A., and Elizabeth B. Strand. 2003. "To Leave or to Stay?" *Journal of Interpersonal Violence* 18 (12): 1367–77. doi:10.1177/0886260503258028.

Fearnley, Lyle. 2008. "Signals Come and Go: Syndromic Surveillance and Styles of Biosecurity." *Environment and Planning A* 40 (7): 1615–32.

Fearnley, Lyle. 2020. *Virulent Zones: Animal Disease and Global Health at China's Pandemic Epicenter.* Durham, NC: Duke University Press.

Ferhani, Adam, and Simon Rushton. 2020. "The International Health Regulations, COVID-19, and Bordering Practices: Who Gets In, What Gets Out, and Who Gets Rescued?" *Contemporary Security Policy* 41 (3): 458–77.

Fine, Paul E. M. 1975. "Ross's a Priori Pathometry—A Perspective." *Proceedings of the Royal Society of Medicine* 68 (9): 547–51.

Fine, Paul E. M. 1979. "John Brownlee and the Measurement of Infectiousness: An Historical Study in Epidemic Theory." *Journal of the Royal Statistical Society. Series A (General)* 142 (3): 347–62. doi:10.2307/2982487.

Finke, Leonhard Ludwig. 1792. *Versuch einer allgemeinen medicinisch-practischen Geographie. Worin der historische Theil der einheimischen Völker- und Staaten-Arzeneykunde Vorgetragen Wird.* Leipzig: Weidmann.

Fiorito, Silvana, Marzia Soligo, Yadong Gao, Ismail Ogulur, Cezmi A. Akdis, and Sergio Bonini. 2022. "Is the Epithelial Barrier Hypothesis the Key to Understanding the Higher Incidence and Excess Mortality during COVID-19 Pandemic? The Case of Northern Italy." *Allergy* 77 (5): 1408–17. doi:10.1111/all.15239.

Flannery, Michael A. 2016. "'Frauds,' 'Filth Parties,' 'Yeast Fads,' and 'Black Boxes': Pellagra and Southern Pride, 1906–2003." *Southern Quarterly* 53 (3): 114–40.

Fleck, Ludwig. 1979 (1935). *Genesis and Development of Scientific Fact.* Translated by F. Bradley and T. Trenn. Edited English translation by F. Bardley and T. J. Trenn. Chicago: University of Chicago Press.

Fleming, Kenneth A., Susan Horton, Michael L. Wilson, Rifat Atun, Kristen DeStigter, John Flanigan, Shahin Sayed, et al. 2021. "The Lancet Commission on Diagnostics: Transforming Access to Diagnostics." *The Lancet* 398 (10315): 1997–2050. doi:10.1016/S0140-6736(21)00673-5.

Flynn, Gerard, and Susan Scutti. 2014. "Smuggled Bushmeat Is Ebola's Back Door to America." *Newsweek Magazine*, August 29. https://www.news week.com/2014/08/29/smuggled-bushmeat-ebolas-back-door-america -265668.html.

Folha de São Paulo. 2020. "Letalidade policial bate recorde, e homicídios sobem durante a pandemia em SP." https://www1.folha.uol.com.br/cotidi ano/2020/07/letalidade-policial-bate-recorde-e-homicidios-durante-a-pan demia-em-sp.shtml.

Fortun, Kim. 2012. "Ethnography in Late Industrialism." *Cultural Anthropology* 27 (3): 446–64. doi:10.1111/j.1548-1360.2012.01153.x.

Foucault, Michel. 1970 (1966). *The Order of Things: An Archaeology of the Human Sciences*. London: Tavistock.

Foucault, Michel. 1973 (1963). *The Birth of the Clinic: An Archaeology of Medical Perception*. New York: Vintage.

Foucault, Michel. 1977. *Discipline and Punish: The Birth of the Prison*. New York: Random House.

Foucault, Michel. 2007. *Security, Territory, Population: Lectures at the Collège de France, 1977–78*. New York: Palgrave.

Foucault, Michel. 2008. *The Birth of Biopolitics: Lectures at the Collège de France, 1978–79*. New York: Palgrave.

Frade, Carlos. 2016. "Social Theory and the Politics of Big Data and Method." *Sociology* 50 (5): 863–77. doi:10.1177/0038038515614186.

Francklyn, Gilbert. 1789. *Observations, Occasioned by the Attempts Made in England to Effect the Abolition of the Slave Trade: Shewing the Manner in which Negroes are Treated in the British Colonies, in the West-Indies; and, also, some Particular Remarks on a Letter Addressed to the Treasurer of the Society for Effecting Such Abolition, from the Rev. Mr. Robert Boucher Nicholls, Dean of Middleham*. Kingston, Jamaica: Reprinted at the Logographic Press, and sold by J. Walter [and two others].

Fressoz, Jean-Baptiste. 2012. *L'apocalypse joyeuse: Une histoire du risque technologique*. Paris: Seuil.

Frontera, Antonio, Claire Martin, Kostantinos Vlachos, and Giovanni Sgubin. 2020. "Regional Air Pollution Persistence Links to COVID19 Infection Zoning." *Journal of Infection* 81 (2): 318–56.

Füller, Henning. 2022. *Infrastrukturen der Biosicherheit*. Bielefeld, Germany: transcript. doi:10.18452/24963.

Galison, Peter. 2010. "Trading with the Enemy." In *Trading Zones and Interactional Expertise: Creating New Kinds of Collaboration*, edited by Michael E. Gorman, 25–52. Cambridge, MA: MIT Press.

Garman, Harrison. 1912. *A Preliminary Study of Kentucky Localities in Which Pellagra Is Prevalent, Having Reference to the Condition of the Corn Crop and to the Possible Presence of an Insect or Other Agent by Which the Disease Spreads*. Bulletin No. 159. Lexington: Kentucky Agricultural Experiment Station

Garnett, Emma. 2020. "Breathing Spaces: Modelling Exposure in Air Pollution Science." *Body & Society* 26 (2): 55–78.

Garrett, Laurie. 2015. "Ebola's Lessons: How the WHO Mishandled the Crisis." *Foreign Affairs* 94: 80.

Gates, Bill. 2018. "Innovation for Pandemics." *New England Journal of Medicine* 378 (22): 2057–60. doi:10.1056/NEJMP1806283.

Gebrekidan, Selam, Katrin Bennhold, Matt Apuzzo, and David D. Kirkpatrick. 2020. "Behind the Curve: Ski, Party, Seed a Pandemic; The Travel Rules That Let COVID-19 Take Flight." *New York Times*, September 30.

https://www.nytimes.com/2020/09/30/world/europe/ski-party-pandemic
-travel-coronavirus.html.

Gentilcore, David. 2013. "'Italic Scurvy,' 'Pellarina,' 'Pellagra': Medical Reactions to a New Disease in Italy, 1770–1815." In *A Medical History of Skin: Scratching the Surface*, edited by Jonathan Reinarz and Kevin Siena, 57–96. London: Pickering & Chatto.

Gentilcore, David. 2015. "Louis Sambon and the Clash of Pellagra Etiologies in Italy and the United States, 1905–14." *Journal of the History of Medicine and Allied Sciences* 71 (1): 19–42.

Gibbes, Philip. 1788. Gibbes to William Fitzherbert, June 9. E20555, Fitzherbert Collection Turner's Hall Plantation, Barbados National Archives, Bridgetown.

Giles-Vernick, Tamara, and James L. A. Webb Jr., eds. 2013. *Global Health in Africa: Historical Perspectives on Disease Control.* Athens: Ohio University Press.

Gleeson, Laura E., Helen M. Roche, and Frederick J. Sheedy. 2021. "Obesity, COVID-19 and Innate Immunometabolism." *British Journal of Nutrition* 125 (6): 628–32. doi:10.1017/S0007114520003529.

Goffman, Erving. 1971. *Relations in Public: Microstudies of the Public Order.* New York: Basic Books.

Goldberger, Joseph. 1922. Joseph Goldberger to Hugh S. Cumming, February 2. Box 153 (1648), Central File, 1897–1923, Record Group 90, Records of the United States Public Health Service, National Archives and Records Administration, College Park, MD.

Goodall, E. W. 1927. "The Epidemic Constitution." *Proceedings of the Royal Society of Medicine* 21 (1): 119–28.

Gostin, Lawrence O., and Eric A. Friedman. 2015. "A Retrospective and Prospective Analysis of the West African Ebola Virus Disease Epidemic: Robust National Health Systems at the Foundation and an Empowered WHO at the Apex." *The Lancet* 385 (9980): 1902–9.

Graham, Janice E. 2019. "Ebola Vaccine Innovation: A Case Study of Pseudoscapes in Global Health." *Critical Public Health* 29 (4): 401–12.

Graham, Loren. 1994. *Science in Russia and the Soviet Union: A Short History.* Cambridge: Cambridge University Press.

Graham, Loren. 2016. *Lysenko's Ghost: Epigenetics and Russia.* Cambridge, MA: Harvard University Press.

Graham, Stephen. 2016. *Vertical: The City from Satellites to Bunkers.* London: Verso Books.

Grasseni, Cristina. 2012. "Community Mapping as Auto-Ethno-Cartography." In *Advances in Visual Methodology*, edited by Sarah Pink, 97–112. Los Angeles: SAGE.

Green, Monica H. 2020. "Emerging Diseases, Re-emerging Histories." *Centaurus* 62 (2): 234–47. doi:10.1111/1600-0498.12306.

Greenhalgh, Trisha, Jose L. Jimenez, Kimberly A. Prather, Zeynep Tufekci, David Fisman, and Robert Schooley. 2021a. "Ten Scientific Reasons in Support of Airborne Transmission of SARS-CoV-2." *The Lancet* 397 (10285): 1603–5.

Greenhalgh, Trish, Mustafa Ozbilgin, and Damien Contandriopoulos. 2021b. "Orthodoxy, *Illusio,* and Playing the Scientific Game: A Bourdieusian Analysis of Infection Control Science in the COVID-19 Pandemic." *Wellcome Open Research* 6 (6): 1–32.

Greenwood, M. 1919. "Sydenham as an Epidemiologist." *Proceedings of the Royal Society of Medicine* 12.Sect. Epidemiol. State Med.: 55–76.

Grimm, Randolph M. 1913. "Pellagra: A Report on Its Epidemiology." *Public Health Reports (1896–1970)* 28 (10) (March 7): 427–50.

Gruetzmacher, Kim, William B. Karesh, John H. Amuasi, Adnan Arshad, Andrew Farlow, Sabine Gabrysch, Jens Jetzkowitz, et al. 2021. "The Berlin Principles on One Health: Bridging Global Health and Conservation." *Science of the Total Environment* 764: 142919. doi:10.1016/j.scitotenv.2020.142919.

Guattari, Félix. 2000. *The Three Ecologies.* London: Athlone Press.

Günther, Thomas, Manja Czech-Sioli, Daniela Indenbirken, Alexis Robitaille, Peter Tenhaken, Martin Exner, Matthias Ottinger, et al. 2020. "SARS-CoV-2 Outbreak Investigation in a German Meat Processing Plant." *EMBO Molecular Medicine* 12 (12): e13296.

Gursky, Elin A., and Gregory Bice. 2012. "Assessing a Decade of Public Health Preparedness: Progress on the Precipice?" *Biosecurity and Bioterrorism: Biodefense Strategy, Practice, and Science* 10 (1): 55–65.

Hacking, Ian. 1994. "Styles of Scientific Thinking or Reasoning: A New Analytical Tool for Historians and Philosophers of the Sciences." In *Trends in the Historiography of Science,* edited by Kostas Gavroglu, Jean Christianidis and Efthymios Nicolaidis, 31–48. Dordecht: Kluwer Academic.

Hamer, William H. 1906. *The Milroy Lectures on Epidemic Diseases in England: The Evidence of Variability and of Persistency of Type Delivered before the Royal College of Physicians of London.* London: Bedford Press.

Hamer, William H. 1928. *Epidemiology Old and New: By Sir William Hamer.* The Anglo-French Library of Medical and Biological Science. London: Kegan Paul, Trench, Trubner.

Hamlin, Christopher. 1998. *Public Health and Social Justice in the Age of Chadwick: Britain, 1800–1854.* Cambridge: Cambridge University Press.

Handler, Jerome S., and Kenneth M. Bilby. 2012. *Enacting Power: The Criminalization of Obeah in the Anglophone Caribbean, 1760–2011.* Kingston, Jamaica: University of the West Indies Press.

Handler, Jerome S., and JoAnn Jacoby. 1993. "Slave Medicine and Plant Use in Barbados." *Journal of the Barbados Museum and Historical Society* 41: 74–98.

Hanemaayer, Ariane. 2019. *The Impossible Clinic: A Critical Sociology of Evidence-Based Medicine.* Vancouver: UBC Press.

Haraway, Donna. 1988. "Situated Knowledges: The Science Question in Feminism and the Privilege of Partial Perspective." *Feminist Studies* 14 (3): 575–99. doi:10.2307/3178066.

Haraway, Donna. 2005. *The Companion Species Manifesto: Dogs, People, and Significant Otherness.* Paradigm, 8. Chicago: Prickly Paradigm Press.

Haraway, Donna. 2015. "Anthropocene, Capitalocene, Plantationocene, Chthulucene: Making Kin." *Environmental Humanities* 6 (1): 159–65.

Haraway, Donna. 2016. *Staying with the Trouble: Making Kin in the Chthulucene.* Durham, NC: Duke University Press.

Hardy, Anne. 2003. "Animals, Disease, and Man: Making Connections." *Perspectives in Biology and Medicine* 46 (2): 200–215.

Harris, Henry F. 1902. "A Case of Ankylostomiasis Presenting the Symptoms of Pellagra." *Transactions of the Medical Association of Georgia* 53: 220–27.

Harrison, Mark. 1999. *Climates & Constitutions: Health, Race, Environment and British Imperialism in India, 1600–1850.* Oxford: Oxford University Press.

Harrison, Mark. 2006. "Disease, Diplomacy and International Commerce: The Origins of International Sanitary Regulation in the Nineteenth Century." *Journal of Global History* 1 (2): 197–217.

Harrison, Mark. 2010. *Medicine in an Age of Commerce and Empire: Britain and Its Tropical Colonies, 1660–1830.* Oxford: Oxford University Press.

Harrison, Mark. 2015. "A Global Perspective: Reframing the History of Health, Medicine, and Disease" *Bulletin of the History of Medicine* 89 (4): 639–89.

Harrison, Mark. 2017. "Pandemics." In *The Routledge History of Disease*, edited by Mark Jackson, 129–46. London: Routledge.

Haushofer, Lisa. 2022. *Wonder Foods: The Science and Commerce of Nutrition.* Vol. 80. Oakland: University of California Press.

Hedeler, Wladislaw, and Meinhard Stark. 2007. "The Grave on the Steppe: The Penal Colony Karaganda in the 1930s." *OstEuropa* 57 (8–9): 589–648.

Heesterbeck, Hans J. A. P. 2002. "A Brief History of R_0 and a Recipe for Its Calculation." *Acta Biotheoretica* 50 (3): 189–204.

Heesterbeck, Hans J. A. P. 2005. "The Law of Mass-Action in Epidemiology: A Historical Perspective." In *Ecological Paradigms Lost*, edited by Kim Cuddington and Beatrix E. Beisner, 81–105. Theoretical Ecology Series. Burlington, VT: Academic Press.

Heffernan, Richard, F. Mostashari, D. Das, M. Besculides, C. Rodriguez, J. Greenko, L. Steiner-Sichel, et al. 2004. "New York City Syndromic Surveillance Systems." *Morbidity and Mortality Weekly Report* 53: 25–27.

Helliwell, Richard, Sujatha Raman, and Carol Morris. 2021. "Environmental Imaginaries and the Environmental Sciences of Antimicrobial Resistance." *Environment and Planning E: Nature and Space* 4 (4): 1346–68.

Henao-Restrepo, Ana Maria, Anton Camacho, Ira M. Longini, Conall H. Watson, W. John Edmunds, Matthias Egger, Miles W. Carroll, et al. 2017. "Efficacy and Effectiveness of an rVSV-vectored Vaccine in Preventing Ebola Virus Disease: Final Results from the Guinea Ring Vaccination, Open-Label, Cluster-Randomised Trial (Ebola Ça Suffit!)." *The Lancet* 389 (10068): 505–18.

Henao-Restrepo, Ana Maria, Marie-Pierre Preziosi, David Wood, Vasee Moorthy, and Marie Paule Kieny. 2016. "On a Path to Accelerate Access to Ebola Vaccines: The WHO's Research and Development Efforts during the 2014–2016 Ebola Epidemic in West Africa." *Current Opinion in Virology* 17: 138–44.

Hendren, Nicholas S., James A. de Lemos, Colby Ayers, Sandeep R. Das, Anjali Rao, Spencer Carter, Anna Rosenblatt, et al. 2021. "Association of Body Mass Index and Age with Morbidity and Mortality in Patients Hospitalized with COVID-19." *Circulation* 143 (2): 135–44. doi:10.1161/ CIRCULATIONAHA.120.051936.

Henning, Kelly J. 2004. "What Is Syndromic Surveillance?" *Morbidity and Mortality Weekly Report* 53: 7–11.

Henning, Kelly J., and Margaret A. Hamburg. 2003. "Syndromic Surveillance." In *Microbial Threats to Health Emergence, Detection, and Response*, edited by Joshua Lederberg, Margaret A. Hamburg, and Mark S. Smolinksi, 281–312. Washington, DC: National Academies Press.

Herrick, Clare. 2019. "Geographic Charisma and the Potential Energy of Ebola." *Sociology of Health & Illness* 41 (8): 1488–502.

Herrick, Clare, and Andrew Brooks. 2020. "Global Health Volunteering, the Ebola Outbreak, and Instrumental Humanitarianisms in Sierra Leone." *Transactions of the Institute of British Geographers* 45 (2): 362–76.

Heymann, David L. 2004. "Smallpox Containment Updated: Considerations for the 21st Century." *International Journal of Infectious Diseases* 8: 15–20.

Hinchliffe, Steve. 2020. "Model Evidence—The COVID-19 Case." Somatosphere, March 31. http://somatosphere.net/forumpost/model-evidence -covid-19/.

Hinchliffe, Steve. 2021. "Postcolonial Global Health, Post-colony Microbes and Antimicrobial Resistance." *Theory, Culture & Society* 39 (3): 145–68. doi:10.1177/0263276420981606.

Hinchliffe, Steve, John Allen, Stephanie Lavau, Nick Bingham, and Simon Carter. 2013. "Biosecurity and the Topologies of Infected Life: From Borderlines to Borderlands." *Transactions of the Institute of British Geographers* 38 (4): 531–43.

Hinchliffe, Steve, Nick Bingham, John Allen, and Simon Carter. 2016. *Pathological Lives: Disease, Space and Biopolitics.* Chichester, UK: John Wiley & Sons.

Hinchliffe, Steve, and Stephanie Lavau. 2013. "Differentiated Circuits: The

Ecologies of Knowing and Securing Life." *Environment and Planning D: Society and Space* 31: 259–74.

Hirsch, August. 1883. *Handbook of Geographical and Historical Pathology.* London: New Sydenham Society.

Hirsch, Lioba A. 2020. "In the Wake: Interpreting Care and Global Health through Black Geographies." *Area* 52 (2): 314–21.

Hobson-West, Pru, and Annemarie Jutel. 2020. "Animals, Veterinarians and the Sociology of Diagnosis." *Sociology of Health & Illness* 42 (2): 393–406. doi:10.1111/1467-9566.13017.

Hoffmann, Markus, Hannah Kleine-Weber, Simon Schroeder, Nadine Krüger, Tanja Herrler, Sandra Erichsen, Tobias S. Schiergens, et al. 2020. "SARS-CoV-2 Cell Entry Depends on ACE2 and TMPRSS2 and Is Blocked by a Clinically Proven Protease Inhibitor." *Cell* 181 (2): 271–80. doi:10.1016/j.cell.2020.02.052.

Hogarth, Rana A. 2017. *Medicalizing Blackness: Making Racial Difference in the Atlantic World, 1780–1840.* Chapel Hill: University of North Carolina Press.

Honigsbaum, Mark. 2016. "'Tipping the Balance': Karl Friedrich Meyer, Latent Infections, and the Birth of Modern Ideas of Disease Ecology." *Journal of the History of Biology* 49 (2): 261–309. doi:10.1007/s10739-015-9430-7.

Honigsbaum, Mark. 2017. "René Dubos, Tuberculosis, and the 'Ecological Facets of Virulence.'" *History and Philosophy of the Life Sciences* 39 (3): 15. doi:10.1007/s40656-017-0142-5.

Honigsbaum, Mark. 2020. *The Pandemic Century: One Hundred Years of Panic, Hysteria, and Hubris.* New York: W. W. Norton.

Honigsbaum, Mark, and Pierre-Olivier Méthot. 2020. "Introduction: Microbes, Networks, Knowledge—Disease Ecology and Emerging Infectious Diseases in Time of COVID-19." *History and Philosophy of the Life Sciences* 42 (3): 1–9.

Hörl, Erich. 2017. "Introduction to General Ecology: The Ecologization of Thinking." In *General Ecology: The New Ecological Paradigm*, edited by Erich Hörl, and James Burton, 1–73. London: Bloomsbury.

Horn, Eva. 2018. "Air as Medium." *Grey Room* 73: 6–25.

House of Commons. 1790. "Minutes, etc., Reported to the House." *House of Commons Sessional Papers of the Eighteenth Century: George III, Minutes of Evidence on the Slave Trade 1790, 71.1, 1975*, edited by Sheila Lambert. Wilmington, DE: Scholarly Resources.

Hsu, Hsuan L. 2020. *The Smell of Risk: Environmental Disparities and Olfactory Aesthetics.* New York: New York University Press.

Huber, Valeska. 2006. "The Unification of the Globe by Disease? The International Sanitary Conferences on Cholera, 1851–1894." *Historical Journal* 49 (2): 453–76.

Hubmann, Michaela. 2021. "Chronicity of Disruptive Project Rhythms: The

Projectification of the 'Post-Ebola' Health System Rebuilding in Sierra Leone." *Time & Society* 30 (3): 379–401.

Immerwahr, Daniel. 2019. *How to Hide an Empire: A History of the Greater United States.* New York: Picador.

Ingold, Tim. 2020. "On Breath and Breathing: A Concluding Comment." *Body & Society* 26 (2): 158–67.

Inniss, Tara. 2009. "'Any Elderly, Sensible, Prudent Woman': The Practice and Practitioners of Midwifery during Slavery in the British Caribbean." In *Health and Medicine in the Circum-Caribbean, 1800–1968*, edited by Juanita De Barros, Steven Palmer, and David Wright, 40–52. Routledge Studies in the Social History of Medicine 33. New York: Routledge.

Instituto Brasileiro de Geografia e Estatística. 2020. "Aglomerados subnormais 2019: Classificação preliminar e informações de saúde para o enfrentamento à COVID-19." Nota Técnica 1. Rio de Janeiro: Instituto Brasileiro de Geografia e Estatística.

Instituto de Estudos Socioeconômicos. 2021. "A conta do desmonte: Balanço do orçamento geral da união 2021." Brasília, Brazil: Instituto de Estudos Socioeconômicos.

Irigaray, Luce. 1999. *The Forgetting of Air in Martin Heidegger.* Austin: University of Texas Press.

Jain, Lochlann. 2007. "Living in Prognosis: Toward an Elegiac Politics." *Representations* 98 (1): 77–92.

James, Peter Bai, Jon Wardle, Amie Steel, and Jon Adams. 2019. "Post-Ebola Psychosocial Experiences and Coping Mechanisms among Ebola Survivors: A Systematic Review." *Tropical Medicine & International Health* 24 (6): 671–91.

Jankovic, Vladimir. 2010. *Confronting the Climate: British Airs and the Making of Environmental Medicine.* New York: Palgrave Macmillan.

Jennings, Allan H., and W. V. King. 1913. "An Intensive Study of Insects as a Possible Etiologic Factor in Pellagra." *American Journal of the Medical Sciences*, ser. 2, 146: 411–40.

Jensen, Nele, Andrew Barry, and Ann H. Kelly. 2021. "More-than-National and Less-than-Global: The Biochemical Infrastructure of Vaccine Manufacturing." *Economy and Society* 52 (1): 9–36.

Jensen, Ole B. 2021. "Pandemic Disruption, Extended Bodies, and Elastic Situations: Reflections on COVID-19 and Mobilities." *Mobilities* 16 (1): 66–80.

Jetelina, Katelyn. 2022. "Understanding Risk." *Your Local Epidemiologist Substack*, March 1. https://yourlocalepidemiologist.substack.com/p/understanding -risk

Jimenez, Jose, Linsey Marr, Shelly Miller, Kimberly Prather, Charles Haas, William Bahnfleth, Richard Corsi, et al. 2021. "FAQs on Protecting Your-

self from COVID-19 Aerosol Transmission." Version 1.88. https://tinyurl
.com/FAQ-aerosols.

Jimenez, Jose, Linsey Marr, Latherine Randall, Edward Thomas Ewing, Zeynep
Tufekci, Trish Greenhalgh, Raymond Tellier, et al. 2022. "What Were the
Historical Reasons for the Resistance to Recognizing Airborne Transmis-
sion during the COVID-19 Pandemic?" *Indoor Air* 32 (8): 32:e13070.

Jones, David, and Stefan Helmreich. 2020. "A History of Herd Immunity." *The
Lancet* 396 (10254): 810–11. doi:10.1016/S0140-6736(20)31924-3.

Jones, James H. 1993. *Bad Blood: The Tuskegee Experiment*. Rev. ed.. New York:
Free Press.

Jones, Susan D. 2004. "Mapping a Zoonotic Disease: Anglo-American Efforts
to Control Bovine Tuberculosis before World War I." *Osiris* 19 (1): 133–48.

Jones, Susan D., and Anna A. Amramina. 2018. "Entangled Histories of Plague
Ecology in Russia and the USSR." *History and Philosophy of the Life Sciences*
40: 49–69.

Jones, Susan D., Bakyt Atshabar, Boris V. Schmid, Marlene Zuk, Anna Am-
ramina, and Nils C. Stenseth. 2019. "Living with Plague: Lessons from the
Soviet Union's Anti-Plague System." *Proceedings of the National Academy of
Sciences U.S.A.* 116 (19): 9155–63.

Jutel, Annemarie. 2009. "Sociology of Diagnosis: A Preliminary Review."
Sociology of Health & Illness 31 (2): 278–99. doi:10.1111/j.1467-9566.2008
.01152.x.

Jutel, Annemarie. 2011. *Putting a Name to It: Diagnosis in Contemporary Society*.
Baltimore: Johns Hopkins University Press.

Jutel, Annemarie, and Sarah Nettleton. 2011. "Towards a Sociology of Diagnosis: Reflections and Opportunities." *Social Science & Medicine* 73 (6):
793–800. doi:10.1016/j.socscimed.2011.07.014.

Kaminski, Hannah, Maël Lemoine, and Thomas Pradeu. 2021. "Immunological Exhaustion: How to Make a Disparate Concept Operational?" *PLoS
Pathogens* 17 (9): e1009892. doi:10.1371/journal.ppat.1009892.

Kang, Min, Jianjian Wei, Jun Yuan, Juxuan Guo, Yingtao Zhang, Jian Hang,
Yabin Qu, et al. 2020. "Probable Evidence of Fecal Aerosol Transmission
of SARS-CoV-2 in a High-Rise Building." *Annals of Internal Medicine* 173
(12): 974–80.

Kassymbekova, Botakoz. 2016. *Despite Cultures: Early Soviet Rule in Tajikistan*.
Pittsburgh: University of Pittsburgh Press.

Kearns, Robin, and Graham Moon. 2002. "From Medical to Health Geography: Novelty, Place and Theory after a Decade of Change." *Progress in
Human Geography* 26 (5): 605–25. doi:10.1191/0309132502ph389oa.

Keck, Frédéric. 2019. "A Genealogy of Animal Diseases and Social Anthropology (1870–2000)." *Medical Anthropology Quarterly* 33 (1): 24–41. doi:10.1111/
maq.12442.

Keck, Frédéric, and Christos Lynteris. 2018. "Zoonosis: Prospects and Challenges for Medical Anthropology." *Medicine Anthropology Theory* 5 (3): 1–14. doi:10.17157/mat.5.3.372.

Keller, Elizabeth Doris. 2022. *Unfit Caretakers: Representations of Enslaved Women and Reproduction in British West Indies Medical Literature, 1764–1833.* Master's thesis, University of Chicago. doi:10.6082/uchicago.4159.

Kelly, Ann H. 2018. "Ebola Vaccines, Evidentary Charisma and the Rise of Global Health Emergency Research." *Economy and Society* 47 (1): 135–61.

Kelly, Ann H., Frédéric Keck, and Christos Lynteris. 2019. *The Anthropology of Epidemics.* London: Taylor & Francis.

Kelly, Ann H., and Javier Lezaun. 2013. "Walking or Waiting? Topologies of the Breeding Ground in Malaria Control." *Science as Culture* 22 (1): 86–107. doi:10.1080/09505431.2013.776368.

Kelly, Ann H., Javier Lezaun, and Alice Street. 2022. "Global Health, Accelerated: Rapid Diagnostics and the Fragile Solidarities of 'Emergency R&D.'" *Economy and Society* 51 (2): 187–210. doi: 10.1080/03085147.2021.2014730.

Kelly, Ann H., and Almudena Marí Sáez. 2018. "Shadowlands and Dark Corners." *Medicine Anthropology Theory* 5 (3).

Kenner, Alison. 2018. *Breathtaking: Asthma Care in a Time of Climate Change.* Minneapolis: University of Minnesota Press.

Keränen, Lisa. 2011. "Concocting Viral Apocalypse: Catastrophic Risk and the Production of Bio(in)security." *Western Journal of Communication* 75 (5): 451–72.

Khalid, Adeeb. 2015. *Making Uzbekistan: Nation, Empire and Revolution in the Early USSR.* Ithaca, NY: Cornell University Press.

Khudyakov, Igor V., and Yuri G. Suchkov. 1999. "On the Anniversary of Klavdiya Aleksandrovna Kuznetsova." In *Illuminating Studies of the Soviet Anti-Plague System*, edited by M.I. Levi. Vol. 9.

Kickbusch, Ilona, and Christian Franz. 2017. "Studie: Die integrierte Umsetzung der Bekämpfung der vernachlässigten Tropenkrankheiten—Potential Deutschlands." *Deutsches Netzwerk gegen vernachlässigte Tropenkrankheiten*, 1–50. https://dntds.de/publikationen.html?file=files/content/pdf/pub likationen/Studie_deu_NTDs_Potential%20Deutschlands_Implemen tierung_2017_fin_.pdf&cid=536.

Kiechle, Melanie A. 2016. "Navigating by Nose: Fresh Air, Stench Nuisance, and the Urban Environment, 1840–1880." *Journal of Urban History* 42 (4): 753–71.

Kiechle, Melanie A. 2017. *Smell Detectives: An Olfactory History of Nineteenth-Century Urban America.* Seattle: University of Washington Press.

Kiechle, Melanie A. 2021. "'Health Is Wealth': Valuing Health in the Nineteenth-Century United States." *Journal of Social History* 54 (3): 775–98.

Kim, Annette M. 2015. "Critical Cartography 2.0: From 'Participatory Mapping' to Authored Visualizations of Power and People." *Landscape and Urban Planning* 142: 215–25. doi:10.1016/j.landurbplan.2015.07.012.

King, Brian. 2017. *States of Disease: Political Environments and Human Health.* Oakland: University of California Press.

King, Nicholas B. 2002. "Security, Disease, Commerce: Ideologies of Postcolonial Global Health." *Social Studies of Science* 32 (5–6): 767–89. doi:10.1177/030631270203200507.

Kirk, Robert G. W., and Michael Worboys. 2013. "Medicine and Species: One Medicine, One History?" In *The Oxford Handbook of the History of Medicine*, edited by Mark Jackson, 561–77. Oxford: Oxford University Press.

Kirksey, Stefan, and Stefan Helmreich. 2010. "The Emergence of Multispecies Ethnography." *Cultural Anthropology* 25 (4): 545–76. doi:10.1111/j.15 48-1360.2010.01069.x.

Kohrt, Brandon A., Amit S. Mistry, Nalini Anand, Blythe Beecroft, and Iman Nuwayhid. 2019. "Health Research in Humanitarian Crises: An Urgent Global Imperative." *BMJ Global Health* 4 (6): e001870.

Korenberg, E. I. 2018. "Anniversary of the Theory of Academician E. N. Pavlovsky about the Natural Focality of Diseases (1939–2014)." *Epidemiology and Vaccinal Prevention* 14 (1): 9–16. doi:10.31631/2073-3046-2015-14-1-9 -16.

Kraemer, Moritz U. G., Robert C. Reiner, Oliver J. Brady, Jane P. Messina, Marius Gilbert, David M. Pigott, Dingdong Yi, et al. 2019. "Past and Future Spread of the Arbovirus Vectors Aedes Aegypti and Aedes Albopictus." *Nature Microbiology* 4 (5): 854–63. doi:10.1038/s41564-019-0376-y.

Kraikovski, Alexei, and Julia Lajus. 2019. "The Metropolitan Bay: Spatial Imaginary of St. Petersburg and Maritime Heritage of the Gulf of Finland." *Humanities* 8 (1): 37. doi:10.3390/h8010037.

Krause, Ute. 2021. "Treponema Pertenue." Pschyrembel online. https://www .pschyrembel.de/Treponema%20pertenue/K0MVL.

Krementsov, Nikolai. 1997. *Stalinist Science.* Princeton, NJ: Princeton University Press.

Krieger, Nancy. 1994. "Epidemiology and the Web of Causation: Has Anyone Seen the Spider?" *Social Science & Medicine* 39 (7): 887–903.

Kristeva, Julia. 1982. *Powers of Horror: An Essay on Abjection.* New York: Columbia University Press.

Krug, Etienne G., James A. Mercy, Linda L. Dahlberg, and Anthony B. Zwi. 2002. "The World Report on Violence and Health." *The Lancet* 360 (9339): 1083–88. doi:10.1016/S0140-6736(02)11133-0.

Kryuchechnikov, V. N. 2000. "Polina Andreevna Petrishcheva (on the 100th Birthday Anniversary)." *Zoologicheskii Zhurnal* 79 (7): 876–80.

KSCQZD (M. Aikimbayev's Kazakh Scientific Center for Quarantine and Zoonotic Diseases, now The Masgut Aikimbayev National Scientific Center of Especially Dangerous Infections). 2010. *Atlas of Bacterial and Virus Zoonotic Infections Distribution in Kazakhstan*. Almaty, Kazakhstan: RPK «Reforma».

Kucharski, Adam. 2020. *The Rules of Contagion: Why Things Spread—And Why They Stop*. New York: Profile Books.

Kuhn, Thomas S. 2012 (1962). *The Structure of Scientific Revolutions*. 4th ed. Chicago: University of Chicago Press.

Lachenal, Guillaume. 2015. "Lessons in Medical Nihilism: Virus Hunters, Neoliberalism, and the AIDS Pandemic in Cameroon." In *Para-States and Medical Science*, 103–41. Durham, NC: Duke University Press.

Lachenal, Guillaume. 2016. "At Home in the Postcolony: Ecology, Empire and Domesticity at the Lamto Field Station, Ivory Coast." *Social Studies of Science* 46 (6): 877–93.

Lahav, Rotem, Yulia Haim, Nikhil S. Bhandarkar, Liron Levin, Vered Chalifa-Caspi, Dylan Sarver, Ageline Sahagun, et al. 2021. "CTRP6 Rapidly Responds to Acute Nutritional Changes, Regulating Adipose Tissue Expansion and Inflammation in Mice." *American Journal of Physiology-Endocrinology and Metabolism* 321 (5): E702–13. doi:10.1152/ajpendo.00299.2021.

Lakoff, Andrew. 2007. "Preparing for the Next Emergency." *Public Culture* 19 (2): 247–71. doi:10.1215/08992363-2006-035.

Lakoff, Andrew. 2012. "Epidemic Intelligence: Toward a Genealogy of Global Health Security." In *Contagion*, edited by Bruce Magnusson and Zahi Zalloua, 44–70. Washington, DC: University Press.

Lakoff, Andrew. 2017. *Unprepared: Global Health in a Time of Emergency*. Berkeley: University of California Press.

Lakoff, Andrew. 2021. "Preparedness Indicators: Measuring the Condition of Global Health Security." *Sociologica* 15 (3): 25–43. doi:10.6092/issn.1971-8853/13604.

Lakoff, Andrew, Stephen J. Collier, and Christopher Kelty. 2015. "Ebola's Ecologies." *Limn* 5. https://limn.it/articles/introduction-ebolas-ecologies/.

Lancaster, Kari, Tim Rhodes, and Marsha Rosengarten. 2020. "Making Evidence and Policy in Public Health Emergencies: Lessons from COVID-19 for Adaptive Evidence-Making and Intervention." *Evidence & Policy* 16 (3): 477–90. doi:10.1332/174426420X15913559981103.

Landecker, Hannah. 2016. "Antibiotic Resistance and the Biology of History." *Body & Society* 22 (4): 19–52. doi:10.1177/1357034X14561341.

Landecker, Hannah. 2024. "How the Social Gets under the Skin: From the Social as Signal to Society as Metabolic Milieu." *KZfSS Kölner Zeitschrift für Soziologie und Sozialpsychologie*. Volume 76: 745–67. doi:10.1007/s11577-024-00951-5.

Lanini, Simone, Alimuddin Zumla, John P. A. Ioannidis, Antonino Di Caro, Sanjeev Krishna, Lawrence Gostin, Enrico Girardi, et al. 2015. "Are Adaptive Randomised Trials or Non-randomised Studies the Best Way to Address the Ebola Outbreak in West Africa?" *The Lancet. Infectious Diseases* 15 (6): 738–45.

Larremore, Daniel B., Bryan Wilder, Evan Lester, Soraya Shehata, James M. Burke, James A. Hay, Milind Tambe, Michael J. Mina, and Roy Parker. 2021. "Test Sensitivity Is Secondary to Frequency and Turnaround Time for COVID-19 Screening." *Science Advances* 7 (1): eabd5393. doi:10.1126/sciadv.abd5393.

Lascelles, Edwin, James Colleton, Edward Drax, Francis Ford, John Brathwaite, John Walter, William Thorpe Holder, et al. 1786. "Instructions for the Management of a Plantation in Barbados and for the Treatment of Negroes." S183.B35 I 5 1786b/141173, West Indies Collection, University of the West Indies, Cave Hill, Barbados.

Latimer, Joanna. 2013. "Being Alongside: Rethinking Relations amongst Different Kinds." *Theory, Culture & Society* 30 (7–8): 77–104.

Latour, Bruno. 1990. "Drawing Things Together." In *Representation in Scientific Practice*, edited by Michael Lynch, and Steve Woolgar, 19–68. Cambridge, MA: MIT Press.

Latour, Bruno. 2004. "Why Has Critique Run Out of Steam? From Matters of Fact to Matters of Concern." *Critical Inquiry* 30 (2): 225–48. doi:10.1086/421123.

Latour, Bruno. 2007. "To Modernize or to Ecologize? That Is the Question." In *Technoscience: The Politics of Interventions*, edited by Kristin Asdal, Brita Brenna and Ingunn Moser, 249–72. Oslo: Unipub.

Latour, Bruno. 2014. "How Better to Register the Agency of Things." Tanner Lectures, held at Yale University, March 26 and 27, 2014. http://www.bruno-latour.fr/sites/default/files/137-YALE-TANNER.pdf.

Latour, Bruno. 2017. *Facing Gaia: Eight Lectures on the New Climatic Regime*. Cambridge: Polity.

Lavinder, Claude H. 1911. "The Salient Epidemiological Features of Pellagra." *Public Health Reports (1896–1970)* 26 (39) (September 29): 1459–68.

Lavinder, Claude H. 1912a. Lavinder to Surgeon-General Rupert Blue, January 29. Box 149 (1648), Central File, 1897–1923, Record Group 90, Records of the United States Public Health Service, National Archives and Records Administration, College Park, MD.

Lavinder, Claude H. 1912b. Lavinder to Surgeon-General Rupert Blue, February 2. Box 149 (1648), Central File, 1897–1923, Record Group 90, Records of the United States Public Health Service, National Archives and Records Administration, College Park, MD.

Lavinder, Claude H. 1912c. Lavinder to L. O. Howard, February 2. Box 149

(1648), Central File, 1897–1923, Record Group 90, Records of the United States Public Health Service, National Archives and Records Administration, College Park, MD.

Lebedenko, A. 1961. *Voina c nevidimim vragom* [War with an invisible enemy]. Moscow: n.p. Copy held by the Wellcome Library, London.

Lederberg, Joshua, Robert E. Shope, and Stanley C. Oaks. 1992. *Emerging Infections: Microbial Threats to Health in the United States.* Washington, DC: National Academy Press.

Lee, Bandy. 2016. "Causes and Cures VIII: Environmental Violence." *Aggression and Violent Behavior* 30: 105–9. doi:10.1016/J.AVB.2016.07.004.

Lee, Bandy. 2019. *Violence: An Interdisciplinary Approach to Causes, Consequences, and Cures.* Oxford: John Wiley & Sons.

Leeuw, Evelyne. 2020. "One Health(y) Cities." *Cities & Health* 5 (suppl1): 26–31. doi:10.1080/23748834.2020.1801114.

Lenton, Timothy M., Sébastien Dutreuil, and Bruno Latour. 2020. "Life on Earth Is Hard to Spot." *Anthropocene Review* 7 (3): 248–72. doi:10.1177/20 53019620918939.

Le Roux, Thomas. 2016a. "Chemistry and Industrial and Environmental Governance in France, 1770–1830." *History of Science* 54 (2): 195–222.

Le Roux, Thomas. 2016b. "Governing the Toxics and the Pollutants: France, Great Britain, 1750–1850." *Endeavor* 40 (2):70–81.

Leslie, Chris. 2002. "'Fighting an Unseen Enemy': The Infectious Paradigm in the Conquest of Pellagra." *Journal of Medical Humanities* 23 (3–4): 187–202.

Levi, Moisey Iosifovich. 1991–1999. *Illuminating Studies of the Soviet Anti-plague System* [series of newsletters in Russian]. Moscow: Self-published. Held by Russian and Commonwealth Independent State Collection, Hoover Institution, Stanford University.

Lewis, Michael D., Julie A. Pavlin, Jay L. Mansfield, Sheilah O'Brien, Louis G. Boomsma, Yevgeniy Elbert, and Patrick W. Kelley. 2002. "Disease Outbreak Detection System Using Syndromic Data in the Greater Washington DC Area." *American Journal of Preventive Medicine* 23 (3): 180–86.

Lezaun, Javier. 2010. "Bioethics and the Risk Regulation of 'Frontier Research': The Case of Gene Therapy." In *Anticipating Risks and Organising Risk Regulation*, edited by Bridget M. Hunter, 208–30. Cambridge: Cambridge University Press.

Li, Bo, Jing Yang, Faming Zhao, Lili Zhi, Xiqian Wang, Lin Liu, Zhaohui Bi, and Yunhe Zhao. 2020. "Prevalence and Impact of Cardiovascular Metabolic Diseases on COVID-19 in China." *Clinical Research in Cardiology* 109 (5): 531–38. doi:10.1007/s00392-020-01626-9.

Li, Yiyan, Xing Yang, and Weian Zhao. 2017. "Emerging Microtechnologies and Automated Systems for Rapid Bacterial Identification and

Antibiotic Susceptibility Testing." *SLAS Technology* 22 (6): 585–608. doi:10.1177/2472630317727519.

Li, Yuguo, Hua Qian, Jian Hang, Xuguang Chen, Pan Cheng, Hong Ling, Shengqi Wang, et al. 2021. "Probable Airborne Transmission of SARS-CoV-2 in a Poorly Ventilated Restaurant." *Building and Environment* 196. doi:10.1016/j.buildenv.2021.107788.

Liboiron, Max. 2021. *Pollution Is Colonialism*. Durham, NC: Duke University Press.

Lincoln, Martha, and Ann N. Sosin. 2022. "Ending Free Covid Tests, the US Policy Is Now 'You Do You.'" *The Nation*, September 9. https://www.thenation.com/article/society/covid-tests-end-pandemic/.

Livingston, Julie. 2005. *Debility and the Moral Imagination in Botswana*. Bloomington: Indiana University Press.

Lock, Margaret M., and Vinh-Kim Nguyen. 2010. *An Anthropology of Biomedicine*. London: Wiley-Blackwell.

Lockhart, Sam M., and Stephen O'Rahilly. 2020. "When Two Pandemics Meet: Why Is Obesity Associated with Increased COVID-19 Mortality?" *Med* 1 (1): 33–42. doi:10.1016/j.medj.2020.06.005.

Logette, Emmanuelle, Charlotte Lorin, Cyrille Favreau, Eugenia Oshurko, Jay S. Coggan, Francesco Casalegno, Mohameth François Sy, et al. 2021. "A Machine-Generated View of the Role of Blood Glucose Levels in the Severity of COVID-19." *Frontiers in Public Health* 9: 1–53. doi:10.3389/fpubh.2021.695139.

Lombardo, Joseph S., Howard Burkom, and Julie Pavlin. 2004. "ESSENCE II and the Framework for Evaluating Syndromic Surveillance Systems." *Morbidity and Mortality Weekly Report* 53: 159–65.

London, Alex John, Olayemi O. Omotade, Michelle M. Mello, and Gerald T. Keusch. 2018. "Ethics of Randomized Trials in a Public Health Emergency." *PLoS Neglected Tropical Diseases* 12 (5): e0006313.

López-Otín, Carlos, and Guido Kroemer. 2021. "Hallmarks of Health." *Cell* 184 (1): 33–63. doi:10.1016/j.cell.2020.11.034.

Lorimer, Jamie. 2017. "Probiotic Environmentalities: Rewilding with Wolves and Worms." *Theory, Culture & Society* 34 (4): 27–48. doi:10.1177%2F0263276417695866.

Louie, Janice K., Meileen Acosta, Michael C. Samuel, Robert Schechter, Duc J. Vugia, Kathleen Harriman, Bela T. Matyas, and the California Pandemic (H1N1) Working Group. 2011. "A Novel Risk Factor for a Novel Virus: Obesity and 2009 Pandemic Influenza A (H1N1)." *Clinical Infectious Diseases* 52 (3): 301–12. doi:10.1093/cid/ciq152.

Lowe, Celia. 2010. "Viral Clouds: Becoming H5N1 in Indonesia." *Cultural Anthropology* 25 (4): 625–49.

Lowe, Henry I. C., Arvilla Payne-Jackson, and Cynthia Johnson, eds. 2009. *The Legacy of African, American & the Caribbean Traditional Medicine: Mind, Body and Spirit.* Kingston, Jamaica: Pelican.

Löwy, Iiana. 1988. "Ludwik Fleck on the Social Construction of Medical Knowledge." *Sociology of Health & Illness* 10 (2): 133–55. doi:10.1111/1467-9566.ep11435448.

Löwy, Iiana. 2021. "Hidden Perils: Diagnosing Asymptomatic Disease Carriers." *Medicine Anthropology Theory* 8 (2): 1–29.

Luke, Timothy W. 1995. "On Environmentality: Geo-power and Eco-knowledge in the Discourses of Contemporary Environmentalism." *Cultural Critique* 31: 57–81.

Luzi, Livio, and Maria Grazia Radaelli. 2020. "Influenza and Obesity: Its Odd Relationship and the Lessons for COVID-19 Pandemic." *Acta Diabetologica* 57: 759–64.

Lwande, Olivia Wesula, Vincent Obanda, Anders Lindström, Clas Ahlm, Magnus Evander, Jonas Näslund, and Göran Bucht. 2020. "Globe-Trotting Aedes aegypti and Aedes albopictus: Risk Factors for Arbovirus Pandemics." *Vector-Borne and Zoonotic Diseases* 20 (2): 71–81. doi:10.1089/vbz.2019.2486.

Lynteris, Christos. 2017. "A 'Suitable Soil': Plague's Urban Breeding Grounds at the Dawn of the Third Pandemic." *Medical History* 61 (3): 343–57. doi:10.1017/mdh.2017.32.

Lynteris, Christos, ed. 2019. *Framing Animals as Epidemic Villains: Histories of Non-human Disease Vectors.* Cham, Switzerland: Palgrave Macmillan.

Lynteris, Christos. 2023. "Afterword: Disease Reservoirs and Spatial Imaginaries in the Time of COVID-19." *Medical Anthropology* 42 (4): 432–36.

Mackenzie, Donald. 2008. *An Engine Not a Camera: How Financial Models Shape Markets.* Cambridge, MA: MIT Press.

Mackey, Tim K., Bryan A. Liang, Raphael Cuomo, Ryan Hafen, Kimberly C. Brouwer, and Daniel E. Lee. 2014. "Emerging and Reemerging Neglected Tropical Diseases: A Review of Key Characteristics, Risk Factors, and the Policy and Innovation Environment." *Clinical Microbiology Reviews* 27 (4): 949–79. doi:10.1128/CMR.00045-14.

MacMahon, Brian, and Thomas F. Pugh. 1971. *Epidemiology: Principles and Methods.* Boston: J. & A. Churchill.

Mahon, Niamh, Beth Clark, Amy Proctor, and Lewis Holloway. 2021. "Exploring Farmers' Understanding of and Responses to Endemic Animal Health and Welfare Issues in the UK." *Veterinary Record* 189 (10): e941. doi:10.1002/vetr.941.

Man, Wing Ho, Wouter A. A. de Steenhuijsen Piters, and Debby Bogaert. 2017. "The Microbiota of the Respiratory Tract: Gatekeeper to Respiratory Health." *Nature Reviews Microbiology* 15 (5): 259–70. doi:10.1038/nrmicro.2017.14.

Manderson, Lenore, and Ayo Wahlberg. 2020. "Chronic Living in a Communicable World." *Medical Anthropology* 39 (5): 428–39. doi:10.1080/01459740.2020.1761352.

Mansfield, James V. 1871a. Mansfield to John Mansfield, August 16. James V. Mansfield Papers, William L. Clements Library, University of Michigan, Ann Arbor.

Mansfield, James V. 1871b. Mansfield to John Mansfield, August 20. James V. Mansfield Papers, William L. Clements Library, University of Michigan, Ann Arbor.

Maricato, Erminia. 2019. "Erminia Maricato: Aos sem-teto, a lei." BrCidades. https://outraspalavras.net/outrasmidias/erminia-maricato-aos-sem-teto-a-lei/.

Marie, Armand. 1910. *Pellagra*. Translated by Claude H. Lavinder and James W. Babcock. Columbia, SC: The State.

Marks, Harry M. 2000. *The Progress of Experiment: Science and Therapeutic Reform in the United States, 1900–1990*. Cambridge: Cambridge University Press.

Marks, Harry M. 2003. "Epidemiologists Explain Pellagra: Gender, Race, and Political Economy in the Work of Edgar Sydenstricker." *Journal of the History of Medicine and Allied Sciences* 58 (1): 34–55.

Marres, Noortje. 2005. "Issues Spark a Public into Being: A Key but Often Forgotten Point of the Lippmann-Dewey Debate." *Making Things Public: Atmospheres of Democracy*, edited by Bruno Latour and Peter Weibel, 208–17. Cambridge, MA: MIT Press.

Massumi, Brian. 2009. "National Enterprise Emergency: Steps toward an Ecology of Powers." *Theory, Culture & Society* 26 (6): 153–85. doi:10.1177%2F0263276409347696.

Mateer, Sean W., Steven Maltby, Ellen Marks, Paul S. Foster, Jay C. Horvat, Philip M. Hansbro, and Simon Keely. 2015. "Potential Mechanisms Regulating Pulmonary Pathology in Inflammatory Bowel Disease." *Journal of Leukocyte Biology* 98 (5): 727–37. doi:10.1189/jlb.3RU1114-563R.

Mbembe, Achille. 2019. *Necropolitics (Theory in Forms)*. Durham, NC: Duke University Press.

Mbembe, Achille, and R. H. Mitsch. 2003. "Life, Sovereignty, and Terror in the Fiction of Amos Tutuola." *Research in African Literatures* 34 (4): 1–26.

McCormack, Derek P. 2018. *Atmospheric Things: On the Allure of Elemental Envelopment*. Durham, NC: Duke University Press.

McInnes, Colin, and Kelley Lee. 2012. *Global Health and International Relations*. Cambridge: Polity Press.

McInnes, Colin, and Simon Rushton. 2013. "HIV/AIDS and Securitization Theory." *European Journal of International Relations* 19 (1): 115–38.

McKendrick, Anderson Gray. 1912. "On Certain Mathematical Aspects of Malaria." *Paludism* 1: 54–66.

McKendrick, Anderson Gray, and M. Kesava Pai. 1912. "XLV.—The R'ate of Multiplication of Micro-organisms: A Mathematical Study." *Proceedings of the Royal Society of Edinburgh* 31: 649–55. doi:10.1017/S03701646000 25426.

McNeill, John R. 2010. *Mosquito Empires: Ecology and War in the Greater Caribbean, 1620–1914*. Cambridge: Cambridge University Press.

McNeill, William H. 1976. *Plagues and Peoples*. Garden City, NY: Anchor Press.

Medzhitov, Ruslan. 2021. "The Spectrum of Inflammatory Responses." *Science* 374 (6571): 1070–75. doi:10.1126/science.abi5200.

Melikishvili, Alexander. 2006. "Genesis of the Anti-plague System: The Tsarist Period." *Critical Reviews in Microbiology* 32: 19–31.

Mendelsohn, J. Andrew. 1998. "From Eradication to Equilibrium: How Epidemics Became Complex after World War I." In *Greater Than the Parts: Holism in Biomedicine, 1920–1950*, edited by Christopher Lawrence and George Weisz, 303–34. Oxford: Oxford University Press.

Mesquita, Felipe Nunes, Karina Serra Silvestre, and Valdir Adilson Steinke. 2017. "Urbanização e degradação ambiental: Análise da ocupação irregular em áreas de proteção permanente na região administrativa de Vicente Pires, DF, utilizando imagens aéreas do ano de 2016." *Revista Brasileira de Geografia Física* 10 (3): 722–34. doi:10.5935/1984-2295.20170047.

Méthot, Pierre-Olivier, and Samuel Alizon. 2014. "What Is a Pathogen? Toward a Process View of Host-Parasite Interactions." *Virulence* 5 (8): 775–85. doi:10.4161/21505594.2014.960726.

Mezes, Carolin. 2024. *Monitoring Pandemic Preparedness: Global Health Security's Politics of Accountability, Development, and Infrastructure*. Frankfurt: Campus.

Michaels, Paula. 2003. *Curative Powers: Medicine and Empire in Stalin's Central Asia*. Pittsburgh: University of Pittsburgh Press.

Miller, Stuart C. 1982. *Benevolent Assimilation: The American Conquest of the Philippines, 1899–1903*. New Haven, CT: Yale University Press.

Milton, Donald K. 2020. "A Rosetta Stone for Understanding Infectious Drops and Aerosols." *Journal of the Pediatric Infectious Disease Society* 9 (4): 413–15.

Mina, Michael J., Roy Parker, and Daniel B. Larremore. 2020. "Rethinking Covid-19 Test Sensitivity—A Strategy for Containment." *New England Journal of Medicine* 383 (22): e120. doi:10.1056/NEJMp2025631.

Mitman, Gregg. 2007. *Breathing Space: How Allergies Shape Our Lives and Landscapes*. New Haven, CT: Yale University Press.

Mitman, Gregg. 2022. "Hotspots, Spillovers and the Shifting Geopolitics of Zoonotic Emerging Infectious Diseases: A Commentary." *International Review of Environmental History* 8 (1): 131–37.

Mitman, Gregg, Michelle Murphy, and Christopher Sellers. 2004. "Introduction: A Cloud over History." *Osiris* 19: 1–17. doi:10.1086/649391.

Mitropoulos, Angela. 2012. *Contract & Contagion: From Biopolitics to Oikonomia*. Wivenhoe, UK: Minor Compositions.

Mol, Annemarie. 2002. *The Body Multiple: Ontology in Medical Practice*. Durham, NC: Duke University Press.

Mol, Annemarie. 2008. *The Logic of Care: Health and the Problem of Patient Choice*. London: Routledge.

Mol, Annemarie, and John Law. 1994. "Regions, Networks and Fluids: Anaemia and Social Topology." *Social Studies of Science* 24 (4): 641–71.

Monmonie, Mark. 2011. "Maps as Graphic Propaganda for Public Health." In *Imagining Illness: Public Health and Visual Culture*, edited by David Serlin, 108–25. Minneapolis: University of Minnesota Press. doi:10.5749/minnesota/9780816648221.003.0006.

Montgomery, Catherine, and Lukas Engelmann. 2020. "Epidemiological Publics? On the Domestication of Modelling in the Era of COVID-19." *Somatosphere*. http://somatosphere.net/2020/epidemiological-publics-on-the-domestication-of-modelling-in-the-era-of-covid-19.html/.

Moon, David. 2013. *The Plow That Broke the Steppes: Agriculture and Environment on Russia's Grasslands, 1700–1914*. New York: Oxford University Press.

Moon, Suerie, Devi Sridhar, Muhammad A. Pate, Ashish K. Jha, Chelsea Clinton, Sophie Delaunay, Valnora Edwin, et al. 2015. "Will Ebola Change the Game? Ten Essential Reforms before the Next Pandemic: The Report of the Harvard-LSHTM Independent Panel on the Global Response to Ebola." *The Lancet* 386 (10009): 2204–21.

Mooney, Graham. 2015. *Intrusive Interventions: Public Health, Domestic Space, and Infectious Disease Surveillance in England, 1840–1914*. Woodbridge, UK: Boydell & Brewer.

Morabia, Alfredo. 2013. *A History of Epidemiologic Methods and Concepts*. Basel, Switzerland: Birkhäuser.

Morawska, Lidia, Joseph Allen, William Bahnfleth, Philomena M. Bluyssen, Atze Boerstra, Giorgio Buonanno, Junji Cao, et al. 2021. "A Paradigm Shift to Combat Indoor Respiratory Infection." *Science* 372 (6543): 689–91.

Morawska, Lidia, and Donald K. Milton. 2020. "It Is Time to Address Airborne Transmission of Coronavirus Disease 2019 (COVID-19)." *Clinical Infectious Diseases* 71 (9): 2311–13.

Morgan, Jennifer L. 2004. *Laboring Women: Reproduction and Gender in New World Slavery*. Early American Studies. Philadelphia: University of Pennsylvania Press.

Morgan, Mary S. 2012. *The World in the Model: How Economists Work and Think*. Cambridge: Cambridge University Press.

Morse, Stephen S. 1992. "Global Microbial Traffic and the Interchange of Disease." *American Journal of Public Health* 82 (10): 1326–27.

Morse, Stephen S. 1996. *Emerging Viruses*. New York: Oxford University Press.

Morse, Stephen S. 2012. "Public Health Surveillance and Infectious Disease Detection." *Biosecurity and Bioterrorism: Biodefense Strategy, Practice, and Science* 10 (1): 6–16.

Morys, Filip, and Alain Dagher. 2021. "Poor Metabolic Health Increases COVID-19-Related Mortality in the UK Biobank Sample." *Frontiers in Endocrinology* 12: 297. doi:10.3389/fendo.2021.652765.

Mukharji, Projit Bihari. 2012. "The 'Cholera Cloud' in the Nineteenth-Century British World. History of an Object-without-an-Essence." *Bulletin of the History of Medicine* 86 (3): 303–32.

Müller, Julian. 2016. *Bestimmbare Unbestimmtheiten. Skizze einer indeterministischen Soziologie*. Paderborn, Germany: Fink.

Murall, Carmen Lía, Kevin S. McCann, and Chris T. Bauch. 2012. "Food Webs in the Human Body: Linking Ecological Theory to Viral Dynamics." *PLoS One* 7 (11): e48812. doi:10.1371/journal.pone.0048812.

Murphy, Michelle. 2006. *Sick Building Syndrome and the Problem of Uncertainty: Environmental Politics, Technoscience, and Women Workers*. Durham, NC: Duke University Press.

Myers, Natasha. 2020. "Becoming Sensor in Sentient Worlds: A More-Than-Natural History of a Black Oak Savannah." *Between Matter and Method*, edited by Gretchen Bakke and Marina Peterson, 73–96. Abingdon, UK: Routledge.

Nações Unidas. 2020. "Relator da ONU diz que Brasil tem que suspender despejos durante pandemia." ONU News, July 3, 2020.

Nading, Alex M. 2013. "Humans, Animals, and Health: From Ecology to Entanglement." *Environment and Society* 4 (1): 60–78.

Nancy, Jean-Luc. 2000. *Being Singular Plural*. Stanford, CA: Stanford University Press.

Narat, Victor, Lys Alcayna-Stevens, Stephanie Rupp, and Tamara Giles-Vernick. 2017. "Rethinking Human–Nonhuman Primate Contact and Pathogenic Disease Spillover." *Ecohealth* 14 (4): 840–50.

Nash, Linda. 2006. *Inescapable Ecologies. A History of Environment, Disease, and Knowledge*. Berkeley: University of California Press.

Nash, Linda. 2014. "Beyond Virgin Soils." In *The Oxford Handbook of Environmental History*, edited by Andrew C. Isenberg, 80–85. Oxford: Oxford University Press.

Nash, Linda. 2015. "Writing Histories of Disease and Environment in the Age of the Anthropocene." *Environmental History* 20 (4): 796–804.

Nash, Linda. 2017. "Beyond Virgin Soils" In *The Oxford Handbook of Environmental History*, edited by Andrew C. Isenberg, 76–107. Oxford: Oxford University Press.

National Academies of Sciences, Engineering, and Medicine. 2022. *Long COVID: Examining Long-Term Health Effects of COVID-19 and Implications for the Social Security Administration: Proceedings of a Workshop.* Washington, DC: National Academies Press. https://doi.org/10.17226/26619.

National Electronic Disease Surveillance System Working Group. 2001. "National Electronic Disease Surveillance System (NEDSS): A Standards-Based Approach to Connect Public Health and Clinical Medicine." *Journal of Public Health Management & Practice* 7: 43–50. doi:10.1097/00124784-20 0107060-00005.

Naunheim, Matthew R., Jonathan Bock, Phillip A. Doucette, Matthew Hoch, Ian Howell, Michael M. Johns, Aaron M. Johnson, et al. 2021. "Safer Singing during the SARS-CoV-2 Pandemic: What We Know and What We Don't." *Journal of Voice* 35 (5): 765–71.

Nguyen, Vinh-Kim. 2019. "An Epidemic of Suspicion—Ebola and Violence in the DRC." *New England Journal of Medicine* 380 (14): 1298–99. doi:10.1056/NEJMP1902682/.

NSSP (National Syndromic Surveillance Program). Centers for Disease Control and Prevention. 2019. *NSSP Update*, September.

Numbers, Ronald L. 2000. "Medical Science before Scientific Medicine: Reflections on the History of Medical Geography." *Medical History Supplement* 44 (S20): 217–20.

Nuñes, João. 2016. "Ebola and the Production of Neglect in Global Health." *Third World Quarterly* 37 (3): 542–56.

O'Brien, Melanie, and Maria Ximena Tolosa. 2016. "The Effect of the 2014 West Africa Ebola Virus Disease Epidemic on Multi-level Violence against Women." *International Journal of Human Rights in Healthcare* 9 (3): 151–60. doi:10.1108/ijrh-09-2015-0027/full/xml.

O'Dempsey, Tim. 2017. "Failing Doctor Kahn." In *The Politics of Fear: Médecins Sans Frontières and the West African Ebola Epidemic*, edited by Michiel Hofman and Sokhieng Au, 175–86. Oxford: Oxford University Press.

OECD (Organization for Economic Co-operation and Development). 2020. "Building Back Better: A Sustainable, Resilient Recovery after COVID-19." *OECD Policy Responses to Coronavirus (COVID-19)*. OECD Publishing, June 5.doi:10.1787/a23f764b-en.

Ohayan, Isabelle. 2006. *La sédentarisation des Kazakhs dans l'URSS de Staline: Collectivisation et changement social (1928–1945)*. Paris: Maisonneuve & Larose.

O'Hearn, Meghan, Junxiu Liu, Frederick Cudhea, Renata Micha, and Dariush Mozaffarian. 2021. "Coronavirus Disease 2019 Hospitalizations Attributable to Cardiometabolic Conditions in the United States: A Comparative

Risk Assessment Analysis." *Journal of the American Heart Association* 10 (5): e019259. doi:10.1161/JAHA.120.019259.

Okumura, Renata. 2022. "Abrigo para pets? Cidade gaúcha atrai moradores de rua ao acolher aambém seus animais." Estadão. https://brasil.estadao.com .br/noticias/geral,para-atrair-morador-em-situacao-de-rua-para-abrigos -no-frio-municipio-do-rs-acolhe-tambem-pets,70004130175.

Oldfield, Jonathan D., and Denis J. B. Shaw. 2015. *The Development of Russian Environmental Thought: Scientific and Geographical Perspectives on the Natural Environment.* Abingdon, UK: Routledge.

Olivarius, Kathryn. 2019. "Immunity, Capital, and Power in Antebellum New Orleans." *American Historical Review* 124 (2): 425–55.

OpGEN. 2021. "OpGen Receives FDA Clearance for Acuitas® AMR Gene Panel Website." https://ir.opgen.com/news-releases/news-release-details/ opgen-receives-fda-clearance-acuitasr-amr-gene-panel.

Opitz, Sven. 2016. "Regulating Epidemic Space: The Nomos of Global Circulation." *Journal of International Relations and Development* 19 (2): 264–84. doi:10.1057/jird.2014.30.

Opitz, Sven. 2020. "Luftsicherheitszonen: Atmosphären des Selbst in Zeiten von COVID-19." In *Die Corona-Gesellschaft. Analysen zur Lage und Perspektiven für die Zukunft,* edited by Michael Volkmer and Karin Werner, 125–34. Bielefeld, Germany: transcript.

Osborne, Michael A. 2000. "The Geographical Imperative in Nineteenth-Century French Medicine." *Medical History Supplement* 44 (S20): 31–50.

Otter, Chris, Nicholas Breyfogle, and John L. Brooke. 2015. "Forum Introduction: Pathogen-Ecology Systems in the Anthropocene." *Environmental History* 20 (4): 711–21.

Pálsson, Gisli. 2013. "Ensembles of Biosocial Relations." In *Biosocial Becomings: Integrating Social and Biological Anthropology,* edited by T. Ingold and P. Gísli, 22–41. Cambridge: Cambridge University Press.

Paludan, Søren R., Thomas Pradeu, Seth L. Masters, and Trine H. Mogensen. 2021. "Constitutive Immune Mechanisms: Mediators of Host Defence and Immune Regulation." *Nature Reviews Immunology* 21 (3): 137–50. doi:10.1038/s41577-020-0391-5.

Parkes, Margot W. 2011. "Diversity, Emergence, Resilience: Guides for a New Generation of Ecohealth Research and Practice." *EcoHealth* 8 (2): 137–39. doi:10.1007/s10393-011-0732-8.

Pastukhov, B. N. 1959. "The Main Organizational and Methodical Principles of the Prophylaxis against Plague in the USSR." *Journal of the Pakistan Medical Association* 9 (8): 10–14.

Paugh, Katherine. 2013. "The Politics of Childbearing in the British Caribbean and the Atlantic World during the Age of Abolition, 1776–1838." *Past & Present* 221 (1): 119–60.

Paugh, Katherine. 2014. "Yaws, Syphilis, Sexuality, and the Circulation of Medical Knowledge in the British Caribbean and the Atlantic World." *Bulletin of the History of Medicine* 88 (2): 225–52.

Paugh, Katherine. 2017. *Politics of Reproduction: Race, Medicine, and Fertility in the Age of Abolition.* The Past & Present Book Series. Oxford: Oxford University Press.

Pauly, Daniel. 1995. "Anecdotes and the Shifting Baseline Syndrome of Fisheries." *TREE* 10 (10): 430.

Pavlovsky, Evgeny Nikanorovich. 1928–1932. "Central Asian Parasitological Expeditions 1928–1932." Fond 878, Opus 1, no. 50, l. 33, Archive of the Russian Academy of Sciences, St. Petersburg Branch.

Pavlovsky, Evgeny Nikanorovich. 1937. "The Pamir Expedition, the Biological Station of the Central Asian State University, and Their Works." n.p., Fond 878, Opus 3, no. 59, l. 23, Archive of the Russian Academy of Sciences, St. Petersburg Branch.

Pavlovsky, Evgeny Nikanorovich. 1939. "On the Natural Foci of Infectious and Parasitic Diseases." *Vestnik Akad. Nauk CCCP* 10: 98–108.

Pavlovsky, Evgeny Nikanorovich. 1942–1943. Pavlovsky's maps and photos of the Iranian expedition. Fond 878, Opus 6, no. 53, l. 75, ARAN-P.

Pavlovsky, Evgeny Nikanorovich. 1946. "Fundamentals of the Theory of Natural Focality of Human Transmissive Diseases." *Zhurnal Obshch Biol.* 67: 3–83.

Pavlovsky, Evgeny Nikanorovich. 1961. "How the Doctrine of the Natural Focus of Diseases in Humans, Animals and Plants Originated and Developed." In Pavlovsky, *General Problems in Parasitology and Zoology*, 5–16. Moscow: Izdatelstvo Akad. Nauka CCCP.

Pavlovsky, Evgeny Nikanorovich. 1966 (1965). *The Natural Focality of Transmissible Diseases.* Translated by Frederick Plous, edited by Norman Levine. Urbana: University of Illinois Press. Originally published as Pavlovsky, Evgeny Nikanorovich. 1965. *Природная Очаговость Трансмиссивных Ъолезней.* Moscow: Izdatelstvo Akad. Nauka.

Peckham, Robert, and Ria Sinha. 2017. "Satellites and the New War on Infection: Tracking Ebola in West Africa." *Geoforum* 80: 24–38.

Pellecchia, Umberto, and Emilie Venables. 2020. "Back to Life—Ebola Survivors in Liberia: From Imaginary Heroes to Political Agents." In *Médecins Sans Frontières and Humanitarian Situations*, 136–53. London: Routledge.

Peters, John Durham. 2015. *The Marvelous Clouds: Toward a Philosophy of Elemental Media.* Chicago: University of Chicago Press.

Peterson, Marina. 2022. *Atmospheric Noise: The Indefinite Urbanism of Los Angeles.* Durham, NC: Duke University Press.

Peterson, Maya. 2019. *Pipe Dreams: Water and Empire in Central Asia's Aral Sea Basin.* Cambridge: Cambridge University Press.

Petrishcheva, Polina Andreevna. 1960. *Razgadannaya Opasnost'* [Unraveling the danger]. Moscow: n.p. A version is available in English: 1962. *Fore-warned Is Forearmed*. Translated by David Myshne. Moscow: Foreign Languages Publishing House.

Petteway, Ryan J. 2022. "On Epidemiology as Racial-Capitalist (Re)Colonization and Epistemic Violence." *Critical Public Health* 30 (1): 1–8. doi:10.108 0/09581596.2022.2107486.

Pfaffenberger, Bryan. 1992. "Social Anthropology of Technology." *Annual Review of Anthropology* 21: 491–516.

Philippopoulos-Mihalopoulos, Andreas. 2016. "Withdrawing from Atmosphere: An Ontology of Air Partitioning and Affective Engineering." *Environment and Planning D: Society and Space* 34 (1): 150–67.

Pichler, Verena, Panayiota Kotsakiozi, Beniamino Caputo, Paola Serini, Adalgisa Caccone, and Alessandra della Torre. 2019. "Complex Interplay of Evolutionary Forces Shaping Population Genomic Structure of Invasive Aedes Albopictus in Southern Europe." *PLoS Neglected Tropical Diseases* 13 (8): e0007554. doi:10.1371/journal.pntd.0007554.

Pickering, Andrew. 2010. *The Mangle of Practice: Time, Agency, and Science*. Chicago: University of Chicago Press.

Pickles, John. 2003. *A History of Spaces: Cartographic Reason, Mapping and the Geo-coded World*. London: Routledge. doi:10.4324/9780203351437.

Pickstone, John V. 2001. *Ways of Knowing: A New History of Science, Technology, and Medicine*. Chicago: University of Chicago Press.

Polianski, Igor J. 2021. "Airborne Infection with Covid-19? A Historical Look at a Current Controversy." *Microbes and Infection* 23 (9–10): 104851.

Prescott, Susan L., Alan C. Logan, Jamie Bristow, Ricardo Rozzi, Rob Moodie, Nicole Redvers, Tari Haahtela, et al. 2022. "Exiting the Anthropocene: Achieving Personal and Planetary Health in the 21st Century." *Allergy* 77 (12): 3498–512. doi:10.1111/all.15419.

Prior, Lucy, David Manley, and Clive E. Sabel. 2018. "Biosocial Health Geography: New 'Exposomic' Geographies of Health and Place." *Progress in Human Geography* 43 (3): 531–52. doi:10.1177/0309132518772644.

Prokhorova, Ninel P. 1972. *Academician E. N. Pavlovsky*. Moscow: Meditsina.

Pyyhtinen, Olli. 2016. *More-Than-Human Sociology: A New Sociological Imagination*. London: Palgrave.

Rajakumar, Kumaravel. 2000. "Pellagra in the United States: A Historical Perspective." *Southern Medical Journal* 93 (3): 272–77.

Ranganathan, Malini, and Carolina Balazs. 2015. "Water Marginalization at the Urban Fringe: Environmental Justice and Urban Political Ecology across the North–South Divide." *Urban Geography* 36 (3): 403–23.

Raposo, Otávio, Ana Rita Alves, Pedro Varela, and Cristina Roldão. 2019. "Negro drama: Racismo, segregação e violência policial nas periferias de Lisboa." *Revista Crítica* 119: 5–28. doi:10.4000/RCCS.8937.

Rehman, Nida. 2022. "Epidemic Infrastructures and the Politics of Responsibility in Lahore." *Antipode* 54 (5): 1451–75. doi:10.1111/anti.12826.

Reingold, Arthur. 2003. "If Syndromic Surveillance Is the Answer, What Is the Question?" *Biosecurity and Bioterrorism: Biodefense Strategy, Practice, and Science* 1 (2): 77–81. doi:10.1089/153871303766275745.

Reubi, David, Clare Herrick, and Tim Brown. 2016. "The Politics of Noncommunicable Diseases in the Global South." *Health & Place* 39: 179–87.

Reverby, Susan L. 2009. *Examining Tuskegee: The Infamous Syphilis Study and Its Legacy*. Charlotte: University of North Carolina Press.

Reyes, Raquel A. G. 2014. "Environmentalist Thinking and the Question of Disease Causation in Late Spanish Philippines." *Journal of the History of Medicine and Allied Sciences* 69 (4): 554–79.

Rhodes, Tim, and Kari Lancaster. 2020. "Mathematical Models as Public Troubles in COVID-19 Infection Control: Following the Numbers." *Health Sociology Review* 29 (2): 177–94. doi:10.1080/14461242.2020.1764376.

Richardson, Eugene. 2020. *Epidemic Illusions: On the Coloniality of Global Public Health*. Cambridge, MA: MIT Press.

Riley, James C. 1987. *Eighteenth Century Campaign to Avoid Disease*. Houndmills, UK: Palgrave Macmillan.

Ring, Natalie J. 2012. *The Problem South: Region, Empire, and the New Liberal State, 1880–1930*. Athens: University of Georgia Press.

Roberts, Stewart R. 1914. *Pellagra: History, Distribution, Diagnosis, Prognosis, Treatment, Etiology*. St. Louis, MO: C. V. Mosby.

Robertson, David Brian. 2015. "The Progressive Era." In *The Oxford Handbook of U.S. Social Policy*, edited by Daniel Béland, Christopher Howard, and Kimberly J. Morgan, 41–58. Oxford: Oxford University Press.

Rock, Melanie, and Patricia Babinec. 2008. "Diabetes in People, Cats, and Dogs: Biomedicine and Manifold Ontologies." *Medical Anthropology* 27 (4): 324–52. doi:10.1080/01459740802427091.

Rodrigues, Carla. 2021. "Utopias atualizadas." In *A força da não violência: Um vínculo etico-político*. São Paulo: Boitempo.

Roe, Daphne A. 1973. *A Plague of Corn: The Social History of Pellagra*. Ithaca, NY: Cornell University Press.

Rogaski, Ruth. 2004. *Hygienic Modernity: Meanings of Health and Disease in Treaty-Port China*. Oakland: University of California Press.

Rohland, Eleanora. 2020. "Covid-19, Climate, and White Supremacy: Multiple Crises or One?" *Journal for the History of Environment and Society* 5: 23–32.

Rolnik, Raquel. 2001. "Territorial Exclusion and Violence: The Case of the State of São Paulo, Brazil." *Geoforum* 32 (4): 471–82. doi:10.1016/S0016-7185(01)00017-3.

Rolnik, Raquel. 2013. "Late Neoliberalism: The Financialization of Homeownership and Housing Rights." *International Journal of Urban and Regional Research* 37 (3): 1058–66. doi:10.1111/1468-2427.12062.

Rolnik, Raquel, and Jeroen Klink. 2011. "Crescimento econômico e desenvolvimento urbano: Por que nossas cidades continuam tão precárias?" *Novos Estudos CEBRAP* 89: 89–109. doi:10.1590/S0101-33002011000100006.

Ronse, Maya, Almudena Marí Sáez, Charlotte Gryseels, Melanie Bannister-Tyrrell, Alexandre Delamou, Alain Guillard, Mustapha Briki, et al. 2018. "What Motivates Ebola Survivors to Donate Plasma during an Emergency Clinical Trial? The Case of Ebola-Tx in Guinea." *PLoS Neglected Tropical Diseases* 12 (10): e0006885.

Rosen, George. 1974. "Economic and Social Policy in the Development of Public Health: An Essay in Interpretation." In Rosen, *From Medical Police to Social Medicine: Essays on the History of Health Care*, 176–200. New York: Science History Publications.

Rosenberg, Charles E. 2002. "The Tyranny of Diagnosis: Specific Entities and Individual Experience." *Milbank Quarterly* 80 (2): 237–60. doi:10.1111/1468-0009.t01-1-00003.

Rosenberg, Charles E. 2012. "Epilogue: Airs, Waters, Places. A Status Report." *Bulletin of the History of Medicine* 86 (4): 661–70. doi:10.1353/bhm.2012.0082.

Ross, Corey. 2017. *Ecology and Power in the Age of Empire: Europe and the Transformation of the Tropical World.* Oxford: Oxford University Press.

Ross, Ronald. 1905a. *Researches on Malaria.* London: John Bale, Sons & Danielsson.

Ross, Ronald. 1905b. "The Logical Basis of the Sanitary Policy of Mosquito Reduction." *Science* 22 (570): 689–99. doi:10.1126/science.22.570.689.

Ross, Ronald. 1916. "An Application of the Theory of Probabilities to the Study of a Priori Pathometry. Part I." *Proceedings of the Royal Society of London A: Mathematical, Physical and Engineering Sciences* 92 (638): 204–30. doi:10.1098/rspa.1916.0007.

Rousell, David. 2021. "A Map You Can Walk Into: Immersive Cartography and the Speculative Potentials of Data." *Qualitative Inquiry* 27 (5): 580–97. doi:10.1177/1077800420935927.

Rupke, Nicolaas A., ed. 2000. *Medical Geography in Historical Perspective.* London: Wellcome Trust Centre for the History of Medicine at UCL.

Rushton, Simon. 2011. "Global Health Security: Security for Whom? Security from What?" *Political Studies* 59 (4): 779–96.

Samavati, Lobelia, and Bruce D. Uhal. 2020. "ACE2, Much More Than Just a Receptor for SARS-COV-2." *Frontiers in Cellular and Infection Microbiology* 10. https://www.frontiersin.org/articles/10.3389/fcimb.2020.00317.

Sambon, Louis W. 1905. "Remarks on the Geographical Distribution and Etiology of Pellagra." *British Medical Journal* 11 (November): 1272–75.

Sánchez, Alexandra, Luciana Simas, Vilma Diuana, and Bernard Larouze. 2020. "COVID-19 nas prisões: Um desafio impossível para a saúde pública?" *Cadernos de Saúde Pública* 36 (5): e00083520. doi:10.1590/0102-311X00083520.

Sanford, Sarah, Jessica Polzer, and Peggy McDonough. 2016. "Preparedness as a Technology of (In)Security: Pandemic Influenza Planning and the Global Biopolitics of Emerging Infectious Disease." *Social Theory & Health* 14 (1): 18–43. doi:10.1057/sth.2015.8.

Sariola, Salla, Roger Jeffery, Amar Jesani, and Gerard Porter. 2019. "How Civil Society Organisations Changed the Regulation of Clinical Trials in India." *Science as Culture* 28 (2): 200–222.

Savitt, Todd L., and James H. Young, eds. 1988. *Disease and Distinctiveness in the American South*. Knoxville: University of Tennessee Press.

Saxinger, Gertrude, Alexis Sancho Reinoso, and Sigrid Irene Wentzel. 2022. "Cartographic Storytelling: Reflecting on Maps through an Ethnographic Application in Siberia." *Fennia-International Journal of Geography* 199 (2): 242–59. doi:10.11143/fennia.110918.

Scheffer, Correspondência M, and Mário Scheffer. 2015. "O capital estrangeiro e a privatização do sistema de saúde Brasileiro." *Cadernos de Saúde Pública* 31 (4): 663–66. doi:0.1590/0102-311XPE010415.

Schiebinger, Londa. 2010. "Theories of Gender and Race." In *Feminist Theory and the Body: A Reader*, edited by Janet Price and Margrit Shildrick, 21–31. Repr., New York: Routledge.

Schiebinger, Londa. 2013. "Medical Experimentation and Race in the Eighteenth-Century Atlantic World." *Social History of Medicine* 26 (3): 364–82.

Schiebinger, Londa. 2017. *Secret Cures of Slaves: People, Plants, and Medicine in the Eighteenth-Century Atlantic World*. Stanford, CA: Stanford University Press.

Schneider, Martin. 2021. "Frambösie." *Pschyrembel online*. https://www.pschyrembel.de/Framb%C3%B6sie/K0868.

Schumacher, Frank. 2016. "Embedded Empire: The United States and Colonialism." *Journal of Modern European History* 14 (2): 202–24.

Scott, James C. 1998. *Seeing Like a State: How Certain Schemes to Improve the Human Condition Have Failed*. New Haven, CT: Yale University Press

Searcy, George H. 1907. "An Epidemic of Acute Pellagra." *Journal of the American Medical Association* 49 (1): 37–38.

Searle, Adam, Jonathon Turnbull, and Jamie Lorimer. 2020. "After the Anthropause: Lockdown Lessons for More-Than-Human Geographies." *Geographical Journal* 187 (1): 69–77. doi:10.1111/geoj.12373.

Seitz, John Britton. 2019. "Science and the Steppe: Agronomists, Nomads and the Settler Colony on the Kazakh Steppe, 1881–1917." PhD diss., Iowa State University.

Sellers, Christopher. 2013. "Health, Work, and Environment: A Hippocratic

Turn in Medical History." In *The Oxford Handbook of the History of Medicine*, edited by Mark Jackson, 450–68. Oxford: Oxford University Press.

Sellers, Christopher. 2018. "To Place or Not to Place: Toward an Environmental History of Modern Medicine." *Bulletin of the History of Medicine* 92 (1): 1–45. doi:10.1353/bhm.2018.0000.

Serres, Michel. 2018. *The Birth of Physics*. London: Rowman & Littlefield.

Seyfert, Robert. 2019. *Beziehungsweisen. Elemente einer relationalen Soziologie.* Weilerswist, Germany: Velbrück.

Shaw, Denis J. B. 2015. "Mastering Nature through Science: Soviet Geographers and the Great Stalin Plan for the Transformation of Nature, 1948–1953." *Slavonic and East European Review* 93 (1): 120–46.

Shaw, Ian Graham Ronald, Paul F. Robbins, and John Paul Jones III. 2010. "A Bug's Life and the Spatial Ontologies of Mosquito Management." *Annals of the Association of American Geographers* 100 (2): 373–92. doi:10.1080/00045601003595446.

Shekiro, Emery, Lily Sussman, and Talia Brown. 2018. "Tracking Drug-Related Overdoses at the Local Level: Using Syndromic Surveillance in the CO-NCR." *Journal of Public Health Informatics* 10 (1). doi:10.5210/ojphi.v10i1.8977.

Shtakelberg, Alexander Alexandrovich. 1930. *Professor Evgeny Pavlovsky, Twenty Years of Scientific Research, 1909–1934.* Moscow: Izdatelstvo Akad. Nauka.

Shukin, Nicole. 2009. *Animal Capital: Rendering Life in Biopolitical Times.* Minneapolis: University of Minnesota Press.

Shuler, Kristrina A. 2011. "Life and Death on a Barbadian Sugar Plantation: Historic and Bioarchaeological Views of Infection and Mortality at Newton Plantation." *International Journal of Osteoarchaeology* 21 (1): 66–81. doi:10.1002/oa.1108.

Siler, Joseph F. 1916. "Medical Notes on Barbados, West Indies. Part Two: Pellagra in Barbados." *American Journal of Tropical Diseases and Preventive Medicine* 3 (4) (October): 168–221.

Silva, Eliana Sousa, Érica Peçanha, and Dalcio Marinho Gonçalves. 2021. *Censo vizinhança USP: Características domiciliares e socioculturais do Jardim São Remo e Sem Terra.* São Paulo: Instituto de Estudos Avançados da Universidade de São Paulo.

Singer, Merill. 2010. "Ecosyndemics: Global Warming and the Coming Plagues of the Twenty-First Century." In *Plagues and Epidemics: Infected Spaces Past and Present*, edited by D. Ann Herring and Alan Swedlund, 21–38. Oxford: Berg.

Singer, Merrill, Nicola Bulled, Bayla Ostrach, and Emily Mendenhall. 2017. "Syndemics and the Biosocial Conception of Health." *The Lancet* 389 (10072): 941–50. doi:10.1016/S0140-6736(17)30003-X.

Singer, R. S., L. A. Hart, and R. L. Zasloff. 1995. "Dilemmas Associated with

Rehousing Homeless People Who Have Companion Animals." *Psychological Reports* 77 (3): 851–57. doi:10.2466/pr0.1995.77.3.851.

Sirleaf, Matiangai. 2018. "Ebola Does Not Fall from the Sky: Structural Violence & International Responsibility." *Vanderbilt Journal of Transnational Law* 51 (2): 477–554.

Sivasundaram, Sujit. 2020. "The Human, the Animal and the Prehistory of COVID-19." *Past & Present* 249 (1): 295–316. doi:10.1093/pastj/gtaa024.

Skloot, Rebecca. 2010. *The Immortal Life of Henrietta Lacks*. New York: Broadway Books.

Sletto, Bjørn Ingmunn. 2009. "'We Drew What We Imagined'—Participatory Mapping, Performance, and the Arts of Landscape Making." *Current Anthropology* 50 (4): 443–76. doi:10.1086/593704.

Sloterdijk, Peter. 2004. *Schäume. Sphären*. Vol. 3, *Plurale Sphärologie*. Frankfurt am Main, Germany: Suhrkamp.

Sloterdijk, Peter. 2011. *Bubbles. Spheres*. Vol. 1, *Microspherology*. Los Angeles: Semiotext(e).

Sneath, David, Martin Holbraad, and Morten Alex Pedersen. 2009. "Technologies of the Imagination: An Introduction." *Ethnos* 74 (1): 5–30. doi:10.1080/00141840902751147.

Snowden, Frank M. 2008. "Emerging and Reemerging Diseases: A Historical Perspective." *Immunological Reviews* 225 (1): 9–26. doi:10.1111/j.1600-065X.2008.00677.x.

Snowden, Frank M. 2020. *Epidemics and Society: From the Black Death to the Present*. New Haven, CT: Yale University Press.

Soper, H. E. 1929. "The Interpretation of Periodicity in Disease Prevalence." *Journal of the Royal Statistical Society* 92 (1): 34–73. doi:10.2307/2341437.

Sotomayor-Bonilla, Jesús, Enrique Del Callejo-Canal, Constantino González-Salazar, Gerardo Suzán, and Christopher R. Stephens. 2021. "Using Data Mining and Network Analysis to Infer Arboviral Dynamics: The Case of Mosquito-Borne Flaviviruses Reported in Mexico." *Insects* 12 (5): 398. doi:10.3390/insects12050398.

SP2. 2022. "Centros de acolhida a moradores de rua na cidade de SP Têm Banheiros entupidos e insetos nos lençóis, aponta relatório." G1 São Paulo. https://g1.globo.com/sp/sao-paulo/noticia/2022/02/18/centros-de-acolhida-a-moradores-de-rua-na-cidade-de-sp-tem-banheiros-entupidos-e-insetos-nos-lencois-aponta-relatorio.ghtml.

Spence, Caroline Q. 2014. "Ameliorating Empire: Slavery and Protection in the British Colonies, 1783–1865." PhD diss., Harvard University.

Spivak, Gayatri. 1998. "Can the Subaltern Speak?" In *Marxism and the Interpretation of Culture*, edited by Cary Nelson and Laurence Grossberk. Hampshire, UK: Macmillan Education.

Stanescu, James. 2012. "Species Trouble: Judith Butler, Mourning, and the

Precarious Lives of Animals." *Hypatia* 27 (3): 567–82. doi:10.1111/j.15 27-2001.2012.01280.x.

Stärk, Katharina D. C., Montserrat Arroyo Kuribreña, Gwenaelle Dauphin, Sandra Vokaty, Michael P. Ward, Barbara Wieland, and Ann Lindberg. 2015. "One Health Surveillance—More Than a Buzz Word?" *Preventive Veterinary Medicine* 120 (1): 124–30. doi:10.1016/j.prevetmed.2015.01.019.

Starks, Tricia. 2008. *The Nody Soviet: Propaganda, Hygiene, and the Revolutionary State.* Madison: University of Wisconsin Press.

Steere-Williams, Jacob. 2019. "'Coolie' Control: State Surveillance and the Labour of Disinfection across the Late Victorian British Empire." In *Transnational Histories*, edited by Robert Heynen and Emily van der Meulen, 35–57. Toronto: University of Toronto Press.

Steere-Williams, Jacob. 2020. *The Filth Disease—Typhoid Fever and the Practices of Epidemiology in Victorian England.* Vol. 49. Woodbridge, UK: Boydell and Brewer.

Stefan, Norbert. 2022. "Metabolic Disorders, COVID-19 and Vaccine-Breakthrough Infections." *Nature Reviews Endocrinology* 18 (2): 75–76. doi:10 .1038/s41574-021-00608-9.

Stefan, Norbert, Andreas L. Birkenfeld, and Matthias B. Schulze. 2021. "Global Pandemics Interconnected—Obesity, Impaired Metabolic Health and COVID-19." *Nature Reviews Endocrinology* 17 (3): 135–49. doi:10.1038/ s41574-020-00462-1.

Stengers, Isabelle. 2005. "Introductory Notes on an Ecology of Practices." *Cultural Studies Review* 11 (1): 183–96. doi: 10.5130/csr.v11i1.3459.

Stengers, Isabelle. 2010. *Cosmopolitics.* Minneapolis: University of Minnesota Press.

Stengers, Isabelle. 2021. "Putting Problematization to the Test of Our Present." *Theory, Culture & Society* 38 (2): 71–92.

Stoto, Michael A. 2012. "The Effectiveness of U.S. Public Health Surveillance Systems for Situational Awareness during the 2009 H1N1 Pandemic: A Retrospective Analysis." *PLoS One* 7: 1–12.

Street, Alice. 2014. *Biomedicine in an Unstable Place: Infrastructure and Personhood in a Papua New Guinean Hospital.* Durham, NC: Duke University Press.

Street, Alice, and Ann H. Kelly. 2021. "Introduction: Diagnostics, Medical Testing, and Value in Medical Anthropology." *Medicine Anthropology Theory* 8 (2): 1–16. doi:10.17157/mat.8.2.6516.

Streule, Monika. 2020. "Doing Mobile Ethnography: Grounded, Situated and Comparative." *Urban Studies* 57 (2): 421–38. doi:10.1177/0042098018817418.

Sutter, Paul S. 2007. "Nature's Agents or Agents of Empire? Entomological Workers and Environmental Change during the Construction of the Panama Canal." *Isis* 98: 724–54.

Swartz, David. 1997. *Culture and Power: The Sociology of Pierre Bourdieu.* Chicago: Chicago University Press.

Swedberg, Richard. 2016. "Before Theory Comes Theorizing or How to Make Social Science More Interesting." *British Journal of Sociology* 67 (1): 5–22.

Sydenham, Thomas. 1676. *Observationes medicae morborum acutorum historiam et curationem.* London: G. Kettilby.

Tartof, Sara Y., Lei Qian, Vennis Hong, Rong Wei, Ron F. Nadjafi, Heidi Fischer, Zhuoxin Li, et al. 2020. "Obesity and Mortality among Patients Diagnosed with COVID-19: Results from an Integrated Health Care Organization." *Annals of Internal Medicine* 173 (10): 773–81. doi:10.7326/M20-3742.

Temkin, Owsei. 2006. "An Historical Analysis of the Concept of Infection." In *The Double Face of Janus and Other Essays in the History of Medicine,* edited by Temkin, 456–71. Baltimore: Johns Hopkins University Press.

Thompson, W. Gilman. 1902. *Practical Dietetics: With Special Reference to Diet in Disease.* New York: Appleton.

Tilley, Helen. 2004. "Ecologies of Complexity: Tropical Environments, African Trypanosomiasis, and the Science of Disease Control in British Colonial Africa, 1900–1940." *Osiris* 19: 21–38. doi:10.1086/649392.

Tsing, Anna Lowenhaupt, Jennifer Deger, Alder Keleman Saxena, and Feifei Zhou. 2021. *Feral Atlas: The More-Than-Human Anthropocene.* Stanford, CA: Stanford Univerity Press. doi:10.21627/2020fa.

Turner, Sasha. 2017. "The Nameless and the Forgotten: Maternal Grief, Sacred Protection, and the Archive of Slavery." *Slavery & Abolition* 38 (2): 232–50.

United Nations. 2020. "COVID-19 and the Right to Adequate Housing: Impacts and the Way Forward." July 1–22.

United Nations. 2021. "11 Goal, Sustainable Cities and Communities. Progress towards the Sustainable Development Goals." E/2021/58. https://unstats.un.org/sdgs/report/2021/goal-11/#.

Upadhya, Vivek. 2013. "Abuse of Animals as a Method of Domestic Violence: The Need for Criminalization." *Emory Law Journal* 23: 1163–209.

Valencius, Convery Bolton. 2000. "Histories of Medical Geography." *Medical History Supplement* 44 (S20): 3–28. doi:10.1017/S0025727300073245.

Valencius, Convery Bolton. 2002. *The Health of the Country: How American Settlers Understood Themselves and Their Land.* New York: Basic Books.

Vasbinder, Alexi, Elizabeth Anderson, Husam Shadid, Hanna Berlin, Michael Pan, Tariq U. Azam, Ibrahim Khaleel, et al. 2022. "Inflammation, Hyperglycemia, and Adverse Outcomes in Individuals with Diabetes Mellitus Hospitalized for COVID-19." *Diabetes Care* 45 (3): 692–700. doi:10.2337/dc21-2102.

Vaughan, Megan. 1991. *Curing Their Ills: Colonial Power and African Illness.* Stanford, CA: Stanford University Press.

Veit, Richard F., Nicola Kelly, Sean McHugh, and Timothy Dismore. 2023. "'Not Unmindful of the Unfortunate': Finding the Forgotten through Archaeology at the Orange Valley Hospital for the Enslaved." *International Journal of Historical Archaeology* 27 (1): 51–80. doi:10.1007/s10761-022-00652-9.

Venables, Emilie. 2017. "'Atomic Bombs' in Monrovia, Liberia: The Identity and Stigmatisation of Ebola Survivors." *Anthropology in Action* 24 (2): 36–43.

Veyne, Paul. 1997. "Foucault Revolutionizes History." In *Foucault and His Interlocutors*, edited by Arnold Ira Davidson, 146–82. Chicago: University of Chicago Press.

Voelkner, Nadine 2011. "Managing Pathogenic Circulation: Human Security and the Migrant Health Assemblage in Thailand." *Security Dialogue* 42 (3): 239–59. doi:10.1177%2F0967010611405393.

Wagner, Michael M., J. Espino, F.-C. Tsui, P. Gesteland, W. Chapman, O. Ivanov, A. Moore, W. Wong, J. Dowling, and J. Hutman. 2004. "Syndrome and Outbreak Detection Using Chief-Complaint Data—Experience of the Real-Time Outbreak and Disease Surveillance Project." *Morbidity and Mortality Weekly Report* 53: 28–31.

Wakefield-Rann, Rachael. 2021. *Life Indoors: How Our Homes Are Shaping Our Bodies and Our Planet*. Singapore: Palgrave Macmillan.

Wald, Priscilla. 2008. *Contagious: Cultures, Carriers, and the Outbreak Narrative*. Durham, NC: Duke University Press.

Wallace, Robert G., Luke Bergmann, Richard Kock, Marius Gilbert, Lenny Hogerwerf, Rodrick Wallace, and Mollie Holmberg. 2015. "The Dawn of Structural One Health: A New Science Tracking Disease Emergence along Circuits of Capital." *Social Science & Medicine* 129: 68–77. doi:10.1016/J .SOCSCIMED.2014.09.047.

Wallace, Rob, Alex Liebman, Luis Fernando Chaves, and Rodrick Wallace. 2020. "COVID-19 and Circuits of Capital." *Monthly Review*, May 1.

Warner, John Harley. 2014. *The Therapeutic Perspective: Medical Practice, Knowledge, and Identity in America, 1820–1885*. Princeton, NJ: Princeton University Press.

Webster, Celestia A. 1867. Diary. Cairns Collection of American Women Writers, Special Collections, University of Wisconsin, Madison.

Weir, Lorna. 2012. "A Genealogy of Global Health Security." *International Political Sociology* 6 (3): 322–25. doi:10.1111/j.1749-5687.2012.00166_4.x.

Weir, Lorna, and Eric Mykhalovskiy. 2010. *Global Public Health Vigilance: Creating a World on Alert*. New York: Routledge.

Welburn, Sue. 2011. "One Health: The 21st Century Challenge." *Veterinary Record* 168 (23): 614–15. doi:10.1136/vr.d3528.

Wenham, Clare. 2019. "The Oversecuritization of Global Health: Changing the Terms of Debate." *International Affairs* 95 (5): 1093–10. doi:10.1093/ ia/iiz170.

Were, Frederick, Brett Archer, Abdourahamane Diallo, Ana Maria Henao Re-
strepo, Alejandro Costa, and Helen Rees. 2019. "Ebola Vaccines: Overview
of the Evidence and Recommendations." Presentation at the Meeting of
the Strategic Advisory Group of Experts (SAGE) on Immunization, Geneva, Switzerland, October. https://terrance.who.int/mediacentre/data/
sage/SAGE_Docs_Ppt_Oct2019/7_session_ebola/Oct2019_Session7_R_
DBlueprint_recommendation.pdf.

Wergin, Carsten. 2023a. *Tourism, Indigeneity, and the Importance of Place: Fighting for Heritage at Australia's Last Frontier.* New York: Lexington Books.

Wergin, Carsten. 2023b. "From Transculture to Transecology: Coming to
Terms with Multispecies Conviviality in the Education for Sustainable De-
velopment." *heiEDUCATION Journal. Transdisciplinary Studies on Teacher
Education* 9: 83–95. doi:10.17885/heiup.heied.2023.9.24724.

Weston, Kath. 2017. *Animate Planet: Making Visceral Sense of Living in a High-
Tech Ecologically Damaged World.* Durham, NC: Duke University Press.

WHA (World Health Assembly). 2020. "'COVID-19 Response' Agenda."
Presented at the Second Plenary Meeting of the WHA, held online, May
18–19, 2020, A73/VR/2.

Whatmore, Sarah. 2006. "Materialist Returns: Practising Cultural Geography in and for a More-Than-Human World." *Cultural Geographies* 13 (4):
600–609.

Whitelaw, Sandy, Anna Baxendale, Carol Bryce, Lindsay MacHardy, Ian
Young, and Emma Witney. 2001. "'Settings' Based Health Approach: A
Review." *Health Promotion International* 16 (4): 339–53.

WHO (World Health Organization). 2005. *International Health Regulations.*
https://www.who.int/publications-detail-redirect/9789241580496.

WHO. 2012. *Bugs, Drugs and Smoke: Stories from Public Health.* https://iris.who
.int/handle/10665/44700.

WHO. 2014. *Ethical Considerations for Use of Unregistered Interventions for
Ebola Viral Disease.* Report of an Advisory Panel to WHO, Geneva.
https://iris.who.int/bitstream/handle/10665/130997/WHO_HIS_KER_
GHE_14.1_eng.pdf?sequence=1.

WHO. 2015. Global Action Plan on AMR. Geneva. http://www.who.int/
antimicrobial-resistance/publications/global-action-plan/en/.

WHO. 2016. "Guidance for Managing Ethical Issues: In Infectious Disease
Outbreaks." http://apps.who.int/iris/bitstream/10665/250580/1/97892415
49837-eng.pdf?ua=.

WHO. 2020a. "Modes of Transmission of Virus Causing COVID-19: Implications
for IPC Precaution Recommendations—Scientific Brief, 27 March 2020."
https://apps.who.int/iris/bitstream/handle/10665/331601/WHO-2019
-nCoV-Sci_Brief-Transmission_modes-2020.1-eng.pdf?sequence=1&
isAllowed=y.

WHO. 2020b. "COVID-19 Is NOT Airborne." *Twitter.* https://twitter.com/WHO/status/1243972193169616898?ref_src=twsrc%5Etfw.

WHO. 2021. "Ebola Vaccines." *Weekly Epidemiological Record* 22 (96) (June 4): 205. https://apps.who.int/iris/bitstream/handle/10665/341623/WER96 22-eng-fre.pdf?sequence=1&isAllowed=y.

WHO. 2022. "UN Environment Programme Joins Alliance to Implement One Health Approach." https://www.who.int/news/item/18-03-2022-un-env ironment-programme-joins-alliance-to-implement-one-health-approach.

WHO. n.d. "Yaws (Endemic Treponematoses)." Accessed January 11, 2023. https://www.who.int/health-topics/yaws#tab=tab_1.

WHO, Department of Immunization, Vaccines and Biologicals. 2016. "Fractional Dose Yellow Fever Vaccine as a Dose-Sparing Option for Outbreak Response." July 20. Secretariat information paper, WHO/YF/SAGE/16.1. https://iris.who.int/bitstream/handle/10665/246236/WHO-YF-SAGE -16.1-eng.pdf?sequence=1.

Whooley, Owen. 2013. *Knowledge in the Time of Cholera: The Struggle over American Medicine in the Nineteenth Century.* Chicago: University of Chicago Press.

Whyte, Susan Reynolds. 2008. "Discrimination: Afterthoughts on Crisis and Chronicity." *Ethnos* 73 (1): 97–100.

Wiegeshoff, Andrea. 2021. "The 'Greatest Traveller of Them All': Rats, Port Cities, and the Plague in U.S. Imperial History (c. 1899–1915)." In *Migrants and the Making of the Urban-Maritime World: Agency and Mobility in Port Cities (c. 1570–1940)*, edited by Christina Reimann and Martin Öhman, 107–26. New York: Routledge.

Wielinga, Peter R. 2013. "Detecting Bioterrorism: How to Detect the Unexpected?" *Biosecurity and Bioterrorism: Biodefense Strategy, Practice, and Science* 11 (S1): 123–24.

Wilkinson, Annie, and Melissa Leach. 2015. "Briefing: Ebola—Myths, Realities, and Structural Violence." *African Affairs* 114 (454): 136–48. doi:10.1093/AFRAF/ADU080.

Winner, Langdon. 1980. "Do Artifacts Have Politics?" *Deadalus* 109: 121–36.

Wolf, Meike. 2015. "Is There Really Such a Thing as 'One Health'? Thinking about a More Than Human World from the Perspective of Cultural Anthropology." *Social Science & Medicine* 129: 5–11. doi:10.1016/j .socscimed.2014.06.018.

Wood, Denis. 1978. "Introducing the Cartography of Reality." In *Humanistic Geography*, edited by David Ley and Marwyn S. Samuels, 206–19. London: Routledge. doi:10.4324/9781315819655.

Woods, Abigail, Michael Bresalier, Angela Cassidy, and Rachel Mason Dentinger. 2017. *Animals and the Shaping of Modern Medicine.* Cham, Switzerland: Palgrave Macmillan.

Worboys, Michael. 2000. *Spreading Germs: Disease Theories and Medical Practice in Britain, 1865–1900*. Cambridge: Cambridge University Press.

Worboys, Michael. 2007. "Was There a Bacteriological Revolution in Late Nineteenth-Century Medicine?" *Studies in History and Philosophy of Science. Part 3: Studies in History and Philosophy of Biological and Biomedical Sciences* 38 (1): 20–42.

Wu, Katherine J. 2023. "The Future of Long Covid." *The Atlantic*, February 13.

Wuhib, Tadesse, Terence L. Chorba, Vladimir Davidiants, William R. Mac Kenzie, and Scott J. N. McNabb. 2002. "Assessment of the Infectious Diseases Surveillance System of the Republic of Armenia: An Example of Surveillance in the Republics of the Former Soviet Union." *BMC Public Health* 2 (1): 1–8.

Yach, Derek, and Douglas Bettcher. 1998. "The Globalization of Public Health: Threats and Opportunities." *American Journal of Public Health* 88 (5): 735–38.

Yong, Ed. 2023. "Long Covid Is Being Erased—Again." *The Atlantic*, April 19.

Yu, Weijun, and Jessica Keralis. 2020. "Controlling COVID-19: The Folly of International Travel Restrictions." *Health and Human Rights*. https://www.hhrjournal.org/2020/04/controlling-covid-19-the-folly-of-international-travel-restrictions/.

Yusoff, Kathryn. 2016. "Anthropogenesis: Origins and Endings in the Anthropocene." *Theory, Culture & Society* 33 (2): 3–28. doi:10.1177/0263276415581021.

Zatta, Marta, Ségolène Brichler, William Vindrios, Giovanna Melica, and Sébastien Gallien. 2023. "Autochthonous Dengue Outbreak, Paris Region, France, September–October 2023." *Emerging Infectious Diseases Journal* 29 (12): 2538–40. doi:10.3201/eid2912.231472.

Zhu, Yongjian, Jingui Xie, Fengming Huang, and Liqing Cao. 2020. "Association between Short-Term Exposure to Air Pollution and COVID-19 Infection: Evidence from China." *Science of the Total Environment* 727: 138704. doi:10.1016/j.scitotenv.2020.138704.

CONTRIBUTORS

OSWALDO SANTOS BAQUERO is Professor Doutor at the University of São Paulo, where he works in the School of Veterinary Medicine and the Institute of Advanced Studies. He coordinates the Multispecies Health Network (Rede de Saúde Multiespécie) at the same university. His formal background comprises veterinary medicine, epidemiology, public health, and data science. Most of his current work has been dedicated to the promotion of multispecies health through praxis concerned with decolonial experience, understanding, and transformation of the health of marginalized multispecies collectives. His research has been published in national and international peer-reviewed journals and books edited in Portuguese.

ULI BEISEL is full professor of human geography at Free University Berlin and currently principal investigator of the international research project Mobile Mosquitoes: Understanding the Entangled Mobilities of *Aedes* Mosquitoes and Humans in India, Mexico, Tanzania and Germany (funded by the Volkswagen Foundation). Her research is situated in the interdisciplinary field of feminist and postcolonial science and technology studies, more-than-human geography, and global health. Uli's work has been published in journals such as *Science as Culture*, *Environmental Humanities*, *Environment and Planning D*, *Geoforum*, and *Medical Anthropology*.

SARA CRISTINA APARECIDA DA SILVA was born and grew in an urban periphery of São Paulo city. She is physiotherapist from the University of São Paulo and currently is conducting a master research on cardiorespiratory physiotherapy at the Heart Institute of the University of São Paulo (Instituto de Coração do Hospital das Clínicas—Incor). During the COVID-19 pandemic, she participated in a community-based research project in an urban periphery of São Paulo city, as well as in a study of Post-COVID-19 Syndrome. She has also volunteered in rescue and adoption of abandoned companion animals.

LUKAS ENGELMANN is a chancellor's fellow and senior lecturer in the history and sociology of biomedicine at the University of Edinburgh. He leads the Epidemy Lab, which is concerned with the history and present of epidemiological reasoning in the twentieth century, funded by an ERC starting grant since 2020. His first book, *Mapping AIDS*, was published with Cambridge University Press in 2018 and considers the visual and medical history of AIDS/HIV. He also published a coauthored monograph with Christos Lynteris, *Sulphuric Utopias* (MIT Press, 2020), which tells the technological history of fumigation and the political history of maritime sanitation at the turn of the twentieth century.

JULIA ENGELSCHALT is a historian and postdoctoral researcher at the Technical University of Darmstadt. She received her PhD from Bielefeld University with a dissertation titled "The Great Obsession: Tropicality in U.S.-American Colonial Medicine and Domestic Public Health, 1898–1924." Her research interests include the entangled histories of medicine, environment, and race; global history; the history of American empire; and, more recently, the history of food infrastructures, nutrition, and hunger in the western Mediterranean. Her publications include chapters in edited volumes published by Routledge and Campus, as well as a coedited volume on knowledge in times of crisis published by transcript.

JÚLIA AMORIM FARIA is a veterinarian and MSc researcher in epidemiology at the University of São Paulo (USP). She is a member of the coordination group of the Multispecies Health Network (Rede de Saúde Multiespécie) and has been involved in community-based research in favelas. In her master's research, she investigates the actions, motivations, difficulties, and possibilities for the articulation of women who live in urban peripheries and work in them protecting nonhuman animals.

HENNING FÜLLER is a researcher at the Geography Department, Humboldt-Universität zu Berlin. With a background in social sciences and political science, he finished his dissertation in 2009 in geography, concerning relational theories of power and urban governance. In current research he engages with the performativity of socio-technical infrastructures and geographies of health. His recently completed habilitation project focused on the role of technologies in governing futures, employing a relational and spatial epistemology. His work has been published in *International Journal of Urban Regional Research*, *Urban Studies*, *Geographical Journal*, and *Erdkunde*. A coedited volume, *Considering Space*, has been released in 2023.

CLARE HERRICK is professor of geography and global health in the Department of Geography at King's College London. Her research concerns the politics and governance of health, with a particular focus on the behavioral and environmental risk factors for chronic disease. Taking a multidisciplinary approach, her research has spanned concerns with the urban politics of obesity prevention, alcohol as a development problem, the commercial determinants of health, the rhetorical and evidentiary politics of global health crises, the complexities of global health partnerships, and the shifting interface between humanitarianism and global health.

STEVE HINCHLIFFE is professor of geography at the University of Exeter. His books include *Pathological Lives* (2017) and *Humans, Animals and Biopolitics: The More Than Human Condition* (2016). He currently works on interdisciplinary projects on disease, biosecurity, and antimicrobial resistant infections, focusing on Europe and Asia. He is a codirector of the Wellcome Centre for Cultures and Environments of Health at Exeter. He sits on the UK government's Scientific Advisory Committee on Exotic Diseases, and he has also been a member of the Department of Environment and Rural Affairs Science Advisory Group's Social Science Expert Group.

SUSAN D. JONES is Distinguished McKnight University Professor in the Program for the History of Science and Technology and the Department of Ecology, Evolution, and Behavior at the University of Minnesota. Her research areas include the history of zoonotic diseases and human-animal interactions; history of ecology and disease; and history of veterinary medicine. Her publications include the books *Death in a Small Package: A Short History of Anthrax* (2010) and *A Concise History of Veterinary Medicine* (with Peter A. Koolmees, 2022). Her current project, *Homelands of the Plagues*, is a monograph about the history of disease ecology and endemic plague in the Soviet Central Asian borderlands.

ANN H. KELLY is Professor of Anthropology at the University of Oxford. She has led multiple transdisciplinary collaborations at the intersections of infectious disease control, health systems strengthening, and emergency R&D, and serves as a member of the WHO Strategic Advisory Group of Experts (SAGE) for Ebola Vaccines and Vaccination. Her ethnographic engagement in those projects has been driven by an abiding concern with the socio-material conditions that structure the production of biomedical knowledge, the local ecologies of labor that circumscribe its circulation and use and the ethical imaginaries that animate collective responses to health crises.

MELANIE A. KIECHLE is associate professor of history at Virginia Tech and the author of *Smell Detectives: An Olfactory History of Nineteenth-Century Urban America* (University of Washington Press, 2017). As a historian of nineteenth-century United States, she focuses on the intersections between science, medicine, lay knowledge, and sensory experience. She has published articles in the *Journal of Social History*, *Journal of Urban History*, *Rethinking History*, *Science as Culture*, and the *Washington Post*'s Made by History column.

HANNAH LANDECKER is a professor of sociology and director of the Institute for Society and Genetics, an interdisciplinary unit at UCLA committed to cultivating research and pedagogy at the interface of the life and human sciences, at the University of California, Los Angeles. She is a historian and sociologist of the life sciences. She holds a joint appointment in the life and social sciences at UCLA. She is the author of *Culturing Life: How Cells Became Technologies* (Harvard University Press, 2007), and has written widely on biotechnology and the intersection of biology and film. Her more recent work concerns the rise of antibiotic resistance, and the history and sociology of metabolism and epigenetics.

CAROLIN MEZES is a sociologist and postdoctoral fellow at University Bielefeld. Her doctorate research was conducted at Philipps-University

Marburg and has been published by Campus with the title *Monitoring Pandemic Preparedness. Global Health Security's Politics of Accountability, Development, and Infrastructure.* Her work has also been published in peer-reviewed journals such as *Leviathan* and *Behemoth*. In her current research she investigates psychosomatic medicine in Germany, building on her interest in science and technology studies and the sociology of medical practice and the psy-disciplines.

SVEN OPITZ is professor of political sociology at Philipps-University Marburg. With a background in social theory, his research addresses the temporal and spatial dimensions in the biopolitics of health security. His most recent work focuses on atmospheric environmentalism, the politics of symbiosis, and the planetary. His research has been published in journals such as *Theory, Culture and Society*, *European Journal of Social Theory*, *Environment and Planning D*, *South Atlantic Quarterly*, *Social Studies of Science*, *Soziale Welt*, *Leviathan or Distinktion*.

KARINA TURMANN is a historian and was a doctoral researcher at the Department of Modern History at Philipps-University Marburg and the Collaborative Research Center's Dynamics of Security (funded by the German Research Foundation). Her work on the instrumentalization of diseases within colonial discourses and practices was published in an edited volume of the Nomos Politics of Security series (2022). In her current research, she focuses on security-related phenomena with a view to German domestic political developments.

CARSTEN WERGIN is an associate professor of social and cultural anthropology at the Heidelberg Center for Transcultural Studies and currently a principal investigator in the international research project Mobile Mosquitoes: Understanding the Entangled Mobilities of *Aedes* Mosquitoes and Humans in India, Mexico, Tanzania and Germany (funded by the Volkswagen Foundation). His work is located at the intersections of heritage, culture, and ecology, with regional foci in Australia, Europe, and the wider Indian Ocean world. He has been published in journals such as the *Anthropological Journal of European Cultures*, *Ethnos*, *Journal of Cultural Economy*, and *Innovation: The European Journal of Social Science Research*.

ANDREA WIEGESHOFF is a historian and affiliated postdoctoral researcher at the University of Marburg. Her research focuses on the history of health and disease, imperial history, and the history of security. Recently, she coedited, with Malte Thießen, a special issue of *Geschichte* *in Wissenschaft und Unterricht* on the history of epidemics (2022) and a special issue, with Benedikt Stuchtey, of the *Journal of Modern European History* on "(In-)Securities across European Empires and Beyond" (2018). Her current book project focuses on the history of epidemics in the nineteenth-century British empire.

INDEX

Note: Page numbers in *italics* indicate figures.

anthropology, 96, 110, 172, 176; cultural, 18; of epidemics, 163

antibiotics, 9; resistance, 120, 127; antibiotic susceptibility testing (AST), 135

Antilles, Lesser, 44, 47

Ashford, Bailey K., 66

Asian tiger mosquito. See *Aedes* mosquitos: *albopictus*

atmosocial condition, 24, 194, 211; atmosocial life in permeable clouds, 194, 201–5, 212; territorial ordering and limits, 194, 195–97; turbulent atmospheres of fluids and affects, 194, 197–201

atmospheric/atmospheres, 193; anxieties about "safe breathing," 200; bodies bathe in, 204; nonrepresentational quality of, 200; renderings, 5–6; shared, 194, 202, 211; space, 193; turbulent, 194, 197–201; volumes, 198

avian influenza, 17

Babcock, James W., 64–65

bacteria(l); foodborne diseases, 16; infections, 129, 136; movement through mucus, 188; nitrogen-fixing, 183; presence in liver, 189; prolongation in air, 200; theory of infection, 63. *See also* pathogenic/pathogens; viruses

Barnet, Clive, 206

Bayer, Ronald, 30

Becker, Norbert, 165–66

"becoming-with" (Haraway), 14

Bennett, Jane, 14

ben Ouagrham-Gormley, Sonia, 38

Benson, Etienne, 7

Berg, Marc, 130

Bignami, Amico, 100

bioinformatics, 126, 137

biomedicine: advancements, 126, 142, 154; biomedical model, limitations, 16, 120, 122; Soviet, 27, 30–32, 40

biopolitics, 16, 23, 144,

biosecurity: and Ebola, 157–58; in livestock farming, 15–16; in US health debates, 115, 117

Blue, Rupert, 65, 67

body-as-ecology, 178, 179. *See also* anthropogenic body-as-ecology

body mass index (BMI), 180

Böhme, Gernot, 199

Bondurant, Eugene, 66–67

Bordier, Arthur, 96

boundedness, 179

Bourouiba, Lydia, 198

Bowker, Geoffrey, 131–32, 134, 138

Brathwaite, John, 54

breath, ecologies of, 193, 197, 198, 224n6; atmospheric anxieties about "safe breathing," 200; breath–affect interconnection, 205; collective breathing, 194; intimacies of shared breath, 202; life in cloudy, 204; social interactions through, 194–95; visual representations of airborne transmission, 201–2

Brownlee, John, 104; work on historical outbreaks, 102–3

bubonic plague, 31, 33, 35, 37, 39–40, 219n20

buildings/built environment, 10, 20, 21, 37, 78–79, 203–4

Buracanã favela, 78–79

Bush, Vannevar, 155

Butler, Judith, 87, 88

Buttimer, Anne, 95

Caines, Clement, 48, 51–53, 56

cardiovascular disease, 180

cartography, 169, 173; immersive, 171; of mosquito distribution, 159–60; research project on human and mosquito mobility, 171. *See also* critical cartography

Casal, Gaspar, 62

Centers for Disease Control and Prevention (CDC), 112; Biosense 2.0 system, 113, 116; about New Jersey Health Department, 118

Chakrabarti, Pratik, 57

Chapin, Charles V., 200, 212

children, enslaved, 54–55; amelioration, 52; neglect of care in plantation system, 51; yaws in, 43, 48, 50, 53, 55–58

chinkungunya, 159, 160

cholera, 97, 162, 207

chronic inflammation, 184; Anthropocene as, 188–90; breakdown at epithelial barriers, 187; in metabolic disorders, 223n5; organismal epithelial-mucosal boundary, 188. *See also* inflammation

classification, 22, 98, 125, 130, 131–32, 134, 138; international disease classification system, 131

climate/climate change, 6, 8, 9, 11–12, 16, 20, 41, 53, 61, 62, 70, 71, 175, 177, 178, 215, 321

clinamen, 198

clinical medicine, emergence of, 12

"cloud(y)" 201

Colgrove, James, 30

collective health, 115, 119, 122

collectivization, 27, 28, 30–31, 35–36, 209

Collier, Stephen, 15

colonialism, 13, 47, 65, 162, 219n3

commensal microbes, 178, 182–83

contact infection, 200

contagion, 50, 61, 102, 192, 204;

"contagionist"/"anti-contagionist" theories of disease transmission, 44–45, 199; and containment, 157; contingent contagionism, 44; temporal horizons of, 143; volatile contagions, 200

Cooper, Melinda, 115, 151

COVID-19 pandemic: air pollution impact, 185; ambient air during, 5–6, 193; Buracanã favela development during, 78, *78*; chronic inflammation and, 187–90; as "dashboard pandemic," 162; discussion of "new normal" during, 210, 211; as epidemic emergency, 214; "etiquette of spacing," 195; following AHA-formula in Germany, 195; forced migration during, 80–81; health security and, 3, 4–6; herd immunity and, 5; impact in urban peripheries, 73, 89; indexical power of, 177; obesity and, 222n1; possibility of infection, 193; rapid tests, 127; structural violence in São Paulo, 81; underlying condition, 179–82, 191

critical cartography, 161; ecological entanglements and, 162; through infrastructural go-along, 169–71, 174–75; possibilities for open-source mapping, 169

Crookshank, Francis G., 107

C-type lectins, 186

Cumming, Hugh S., 71

Davletov, Mengli, 32

DDT (dichlorodiphenyltrichloroethane), 39, 41

Defense Advanced Research Projects Agency (DARPA), 115

dementia, 62, 220n2

Democratic Republic of the Congo (DRC), 147, 149–50

dengue, 23, 159–60

diabetes, 177, 180–81, 184, 187

diagnosis, 126, 129–30; cognitive approaches, 130; etymology of, 129; Fleck's work, 130; Foucault's work, 131; issues in social scientific engagements, 132; process in animal health sector, 128–29; Rosenberg's "The Tyranny of Diagnosis," 133–34; securing epidemic ecologies and, 137–38; sociological approaches, 130, 132; space-times of, 134; "telling apart," 129–30; variation in logics and practices, 135

diagnostic machine, 22, 125–26, 135–37

diagnostic tests, 126, 127–28, 135

disability, 86, 215

disease causation, 21, 61, 72

disease ecology, 15, 19, 32, 61, 70, 110-11, 120, 151, 204

disease eradication 9, 39, 146, 208; campaigns, 40–41

disease reservoir, 143, 151–53. *See also* regulatory frontiers; vaccine/vaccination: "ring vaccination" strategy

disinfection, 35, 39; labor of, 39

Drake, Daniel, 96, 97

droplets, 196–97, 202

Droumeva, Milena, 173

Durkheim, Émile, 130

Dutreuil, Sébastien, 223n7

Ebola, 74; Ad26.ZEBOV/MVA-BN-Filo vaccine, 149, 150; biosecurities, 157; Ebola ça Suffit trial, 146–48, 157; experimental therapies for, 145; natural reservoir, 143; "necropolitics" of Ebola response, 151; outbreaks in Africa, 141, 213; preparedness for, 149; R&D response, 144; rapid tests, 127; rVSV-ZEBOV-GP vaccine, 146, 147, 149, 156; spatial rationales, 23; survivorship, 152–53, 213; vaccine development, 142–43

ECDC. *See* European Centre for Disease Prevention and Control

ecological/ecology, 7; body of infection, 190–91; conceptual framework, 14–15, 16; entanglements, 162; human movement, 17; in-host ecology, 178, 179; niche, 15; modes of inquiry, 14–15; network ecology, 161; of practices, 14; reasoning, 6–7; relationism in disease control, 18; renderings of infectious diseases, 9; reorientation, 14; "styles of reasoning," 8; thinking, 9, 11, 16, 69, 70, 94, 212

ecologies of violence, 21, 84–85, 212; families and household environments, 83–84; forced displacement/migration and, 80–81; during health emergencies, 76–77, 85–86, 89; incarceration and, 81–83; land occupation and, 78–80; problem of syndemics in favela, Brazil, 74–76; unhoused multispecies collectives and, 81; urban capitalism and, 77–78

effluvia, 199

EID. *See* emerging infectious diseases

emergency: COVID-19, 75, 78, 141; Ebola, 74, 141, 152; emergency exceptionalism, 144; emergency R&D, 23, 154–57, 213, 214 emerging infectious diseases (EID), 8–10, 208; collaborative efforts at health security, 10–11; emerging diseases worldview, 9, 110, 115, 120; and health emergencies, 74; HIV/AIDS, 9; sanitation programs, 12. *See also specific entries*

empire: American, 64, 66; British, 46, 48, 209

emission, 6

emulsifiers, 179, 187, 188, 189

encephalitis viruses, 34, 36, 40

entomology, 18, 31, 32, 65, 67, 69, 101, 160, 161, 162, 165, 167

environment(al), 12, 14, 15, 17–18, 47; causes population density of mosquitoes, 101; climate impact on etiology of pellagra, 70; damage, 87; determinism, 108; in epidemiological modeling, 94; factors causes pellagra, 66; "harmful environments" scenario, 57; Hippocratics' consideration of, 95–96; history, 13; infrastructuring, 163; role in infectious diseases, 50–51, 54; as subject of discrete scientific projects, 99; violence, 87

environmentality, 14, 17–18; probiotic, 17–18

epidemic(s); Brownlee's work on epidemic outbreaks, 102–4; constitution, 95, 106–8; discourses of epidemic outbreaks, 87–88; epidemic ecologies, securing, 137–38, 211, 212, 215; frontiers, 155–56; Guldberg and Waage's mass-action principle, 103; Hamer's work on epidemic outbreaks, 104–5; modeling (*see* model/modeling); McKendrick's

regulation of epidemic curves, 105; Ross's work on malaria transmission, 100–102; Soper's epidemic theory, 105–6

epidemic theory: epidemic wave theory, 106–7; McKendrick's principle of, 105, 106; shift in, 21; Soper's principle of, 105, 106

epidemiology, 98–99, 181; imperial, 95; mathematical, 100, 103, 106, 108; of pellagra in United States, 60–61; of Soviet Union, 31–32, 41–42; Steere-Williams' work in, 98. *See also specific entries*

Epidemiology: Old and New (Hamer), 106

epithelial barrier hypothesis, 187, 189, 212

epizootic disease, 125

Ervebo. *See* rVSV-ZEBOV-GP vaccine

ethnography, 21, 23, 162; mobile ethnographies, 172–73, 174; multispecies ethnography, 163. *See also* "infrastructural go-along" method

etiology: *vs.* epidemiology, 60–61; multifactorial, 120; of pellagra, 60, 62–71

European Centre for Disease Prevention and Control (ECDC), 161, 164; cartographic powers of, 171; mapping *A. albopictus* movement, 165; tracking spread of *"Aedes* invasive mosquitoes," 166, 167

Everts, Jonathan, 162

evolution, 138: epidemic, 107; microbial, 125; pathogenic, 4; viral, 185, 187, 190

Fairchild, Amy L., 30

Fanzago, Francesco, 62

Farr, William, 102
favela, 74–75, 77, 212, 215; Buracanã, 78, *79*, 80; São Paulo, 21, 75, 89; São Remo, 75, *78*, 83
FDA. *See* US Food and Drug Administration
Feral Atlas (Tsing), 163
Fine, Paul E. M., 102, 103, 221n4
Finke, Leonhard Ludwig, 96
Finlay, Carlos, 65
Fleck, Ludwig, 130
Food and Agriculture Organization (FAO), 86
The Forgetting of Air in Martin Heidegger (Irigaray), 203
Foucault, Michel: biopolitics, 16; *Birth of the Clinic*, 131; "environmental" security dispositif, 16; environmentality, 14, 15, 17–18; social establishment of disease knowledge, 131
Frade, Carlos, 119
frambesia tropica. See yaws
Francklyn, Gilbert, 51–52
frontiers, 144, 153–58

Galison, Peter, 205
Gamaleya, N. G., 218n11
Garman, Harrison, 67–68
Garnett, Emma, 204
general ecology, 4, 15–16
geography, 95, 106, 178; of health, 121; human, 18; medical, 96–97, 121
germ theory, 60, 62–63, 98–99
Ghebreyesus, Tedros Adhanom, 147
Gibbes, Philip, 51
global health, 114, 121, 123; challenges, 159, 164, 169; community, 141–42; debate on One Health, 120; frontiers of, 154–55; Global Health debate,

123; humanity and, 175; impacting factors, 159; innovation, 142, 156; security/securitization of, 5, 6, 8–10, 114
glucose, 177, 180, 181, 184, 186–87, 189, 191, 222n1
Goffman, Erving, 194–95, 205
Goldberger, Joseph, 60, 71
Goodall, E. W., 107
Gorgas, William C., 65, 108
Graham, Loren, 218n12
Greenwood, M., 107
Grimm, Randolph M., 67
Guldberg, Maximilian, 103

Hamer, William, 103; "theory of epidemic wave," 106–7; work on epidemic outbreaks, 104–5
Handbook of Geographical and Historical Pathology (Hirsch), 96
Haraway, Donna, 14
Harris, Henry F., 62
Harrison, Mark, 10
health emergencies, 73; Ebola and, 74; ecologies of violence during, 76–77, 85–86, 89; hegemonic definitions of violence, 88
health security, 5, 6, 8–10, 114, 178, 207, 214; atmospheric renderings, 5–6; COVID-19 pandemic and, 3; disease threats, 17; ecological approaches to, 208, 14–15; environmentality, 14, 15; through epidemic ecologies, 213; human-animal entanglements/relations, 4–5; population topologies, 5; Soviet perspective on, 41–42
Heesterbeck, Hans J. A. P, 103–6
herd immunity, 5, 146
Hippocratic corpus, 11, 13, 95–96, 212

Hirsch, August, 96–98
HIV/AIDS, 9, 214; disease eradication, 208; zoonotic origins of, 151
hookworm disease, 70
humanitarian aid/need, 23, 36–37, 65, 142, 144, 147, 149, 156
humanitarian experimentation, 153
hypertension, 177, 180–81

immunity, 99, 177; herd, 5, 146
immune system: elevated glucose levels affecting, 186; innate, 182, 189, 223n3; nonspecific, 223n3; role in human bodies, 212
incarceration, violence in, 81–83
industrialization, 12, 28, 40, 177, 223n4; diets, 23
Inescapable Ecologies (Nash), 59
infectious disease, 6, 9, 12, 13, 17, 18, 21, 37, 41, 42, 47, 63, 76, 77, 88, 98, 100, 102, 105, 114, 125, 135, 159, 172, 193, 213, 221n3; "indexical power" of, 176–77. *See also* emerging infectious diseases (EID); model/modeling
inflammation, 177, 181, 184, 187–90, 223n5; physiological, 190. *See also* chronic inflammation
influenza pandemic, 107, 108; H1N1 outbreak, 112, 182; H5N1 influenza virus, 201
infrastructure, 5, 7, 10, 22, 28, 34, 37, 66, 163, 164, 204; infrastructural ecology, 12; infrastructural go-along (*see* "infrastructural go-along" method); infrastructural neglect, 168; infrastructuring, 163; mobility/transportation, 23, 161; public health, 7, 37, 70, 110; socio-technical, 110, 116; surveillance, 41, 109,

168. *See also* "infrastructural go-along" method; surveillance; syndromic surveillance in United States
"infrastructural go-along" method, 23, 164, 169–71, 172–73, 174–75
insect(s), 32–36, 39, 41, 44, 61, 63, 65–66, 72, 80, 101, 165; theory of Garman, 67–68
International Health Regulations, 164–65
International Sanitary Conference (1851), 10
invasive species/invasion, 23; *A. albopictus*, 164, 166–67; linear invasion narrative, 165; and linear progression, 161, 163, 164; Rehman's observation on *A. albopictus*, 168–69
Irigaray, Luce, 203

Jain, Loclann, 153
Jennings, Allan, 68–69
Johnson & Johnson (J&J), 149; Ad26.ZEBOV/MVA-BN-Filo vaccine, 149, 150

Kara-Kirghiz Autonomous Oblast, 34
Karaganda Corrective Labor Camp (Karlag), 28
Kazakhstan, 28, 29, 30, 34-35, 36, 37-38
Kazakh steppes, 29, 218n9; 1916 rebellion among, 28–29; Soviet scientists' disease control activities in, 40, 209
Keller, Elizabeth Doris, 46–47
Kelty, Christopher, 15
King, W. V., 68
Kivu Ebola epidemic, 147
Koch, Robert, 62

Parsson, Gísli, 122

Pasteur, Louis, 62

pathogenic/pathogens, 211–12; airborne transmission, 193; animals as carriers of, 12; causes emerging infectious diseases, 74; harbourer of, 12; multispecies studies of entanglements, 163; pathogenic agents, 82–83; relationship to natural milieu, 15; zoonotic spillover of, 9, 87–89, 144. *See also* bacteria(l); viruses

pathology: geographical, 95–96; Hirsch's works in, 97–98; historical, 97–98; humoral, 220n7; milestones in, 96–97

Paugh, Katherine, 46

Pavlovsky, Evgeny Nikanorovich, 32, *33*, 33-37, 41, 218n15

pellagra research 20, 64–69, 70–71, 209; debates over etiology and epidemiology, 59, 60–61, 71–72; disease control practices in colonies, 65–66; ecological framework, 72; etiologies in Europe before 1900, 62–64; microbial etiologies, 220n1; nutritional origin, 60; pellagra as "disease of place," 70, 71. *See also* vector hypothesis/theory in pellagra

Pellecchia, Umberto, 153

periphery, 21, 75, 99

permeable clouds, atmosocial life in, 201–2; intimacies of shared breath, 202–3; link between atmospheres and habitation, 203–4; visual representations of airborne transmission, 201–2, *202*

Peters, John Durham, 201

Petrishcheva, Polina Andreevna, 32–33, *33*, 36–37, 218n11

physical distance, 195, 200

physical precondition theory, 51

place: neutrality, 59–60, 72; pellagra as "disease of place," 70, 71; role in construction of expertise, 209

plague. *See* bubonic plague

planetary health, 6, 9, 22, 178

plantation colonies: abolitionism and, 48–49; disease control, 49–52, 53–54, 58; lack of medical facilities in plantation, 55–56; reproduction for workforce, 47–49, 54–55; vulnerability of children, 43–44, 56, 58. *See also* yaws

population(s), 5, 7, 16–17, 32, 35; disease in, 120, 157, 182, 211; dynamics, 101–5; exploited, 10; global, 126; host, 94; incarcerated, 81; pathogens and, 107; poor rural, 62, 66, 71; security, 5; surveillance, 38, 167; topologies, 4–6; urban, 74, 77–81; vaccinating, 146; vulnerable, 127, 146

"political ecology of things," 14

preparedness 9, 24, 74, 110, 117, 143–45, 149, 154, 208 public health, 21, 93, 110; apparatus, 30; authorities/experts/officials, 17, 20, 29, 35, 40, 60, 64, 65, 66, 147, 192, 209, 210, 212; care, 23; danger or threats, 9, 116, 123; emergency (*see* emergency); as environmental problem, 116, 119; infrastructure, 7, 37, 70, 109; institutions, 193, 196, 205; maps, 23; "new public health," 99; as political/issue of security, 114–15, 119; preparedness, 117; surveillance/monitoring, 109, 111–15, 118, 123. *See also* emergency: global health; health emergencies; health security;

132, 175, 178, 188, 212, 214–15.
See also biosecurity; COVID-19
pandemic: health security and;
emerging infectious diseases
(EID): collaborative efforts at
health security; global health:
security/securitization of; health
security; population: security;
preparedness
security dispositif, 16
securitization, 4–8, 22, 93, 110, 115,
151, 178; Copenhagen School, 7.
See also security
Seeing Like a State (Scott), 30
Sellers, Christopher, 13, 59–60, 72
senses/sensing: of insecurity, 31, 38;
remote sensing technologies, 17;
societal, 118
September 11 (9/11) attack in United
States, 109, 114–15
Serres, Michel, 198
Shaw, Denis J. B., 167–68
simulation: computer simulation of
indoor airflows, 204; of possible
pandemic futures, 5; with tracer
gas, 196, 203
Shope, Robert, 114
Shukin, Nicole, 122
Siler, Joseph F., 65, 68–69
Simulium fly (sand fly), 63, 72
skin (disease), 46, 49–50, 52–53, 55,
57
slavery, 11, 50, 85; amelioration in
West Indian plantations, 48,
51–53, 56; ban of slave trade,
219–20n5; midwifery during,
220n8. *See also* abolitionism in
West Indian plantation system;
motherhood; plantation colonies
smallpox, 16, 102, 143; eradication
of, 9, 146; inoculation of, 56
Snow, John, 162

social determinants of health, 177
social ordering, 195; disruption of,
214–15
social relations, 211–13
Soper, Herbert Edward, 105–6
Sorting Things Out (Bowker and
Star), 131
The Sources and Modes of Infection
(Chapin), 200
Soviet Union: biological weapons
program, 219n20; collectivization, 27, 28, 30–31, 35–36, 209;
ecological approach to epidemiology, 31–32; erasing nomadism
on steppe peoples, 28–29; securing Central Asian steppe against
disease, 30; Virgin Lands campaign (1954–1960), 41. *See also*
surveillance: disease surveillance
in Soviet Union
space, 134, 193, 222n1; atmospheric
notions of disease space, 6; geographical, 101, 106; topographical, 94, 100–101, 106
spillover: avoiding further viral,
between animals and humans, 4,
9, 13, 74, 89, 144, 151; cross-
species interactions, 120
stable fly (*Stomoxys calcitrans*), 68
Stalin, Josef, 27
standard/standardization: of living,
72; practices/systems/diagnos-
tics, 105, 126, 132, 133, 135,
138, 143, 144, 154, 156; RCTs,
145–46, 147; samples, 127
Stanescu, James, 88
Star, Susan Leigh, 131–32, 134, 138
statistics, 69, 70, 98, 106, 112, 180,
181
Steere-Williams, Jacob, 39, 95, 98
Stengers, Isabelle, 14
Stiles, Charles W., 66